AMIR D. ACZEL

Probability 1

The book that proves there is life in outer space

A HARVEST BOOK
HARCOURT, INC.
San Diego New York London

Library of Congress Cataloging-in-Publication Data
Aczel, Amir D.
Probability 1/Amir D. Aczel.
p. cm.
Includes bibliographical references and index.
ISBN 0-15-100376-9
ISBN 0-15-601080-1 (pbk.)
1. Life on other planets. 2. Cosmology. 3. Mathematics. 4. Biology.
I. Title.
QB54.A25 1998
576.8'39—dc21 98-16868

Text set in Electra
Designed by Linda Lockowitz
Printed in the United States of America
First Harvest edition 2000
F E D C B A

For Debra

Praise for **Probability 1**

"There are two reasons for recommending [*Probability 1*] to any person interested in the debate: It is clearly and gracefully written and it is up-to-date in its astronomical data."

—Martin Gardner, *Los Angeles Times Book Review*

"For one practiced in dealing with numbers, Amir Aczel certainly has a way with words. . . . Whether or not you are ultimately convinced, his fusion of math with delightful prose, all peppered with amusing anecdotes and intriguing history, will make the read worthwhile."

—Jeanette Brown, *Astronomy*

"Statistics are extremely powerful and important, and Aczel is a very clear and capable exponent of them."

—*The New York Times Book Review*

""[Aczel's] powers of persuasion owe less to any statistically certain accident, however, than to his skillful presentation of complex ideas."

—*The New Yorker*

"The argument is wholly convincing, and Aczel presents it in absorbing style, taking us through the science and mathematics with commendable lucidity."

—*The Financial Times* (London)

"*Probability 1* is a clear and accessible account of the prospects for the existence of intelligent life in the Universe. Amir Aczel covers a lot of ground at a level anyone interested in the puzzle of our origins will find easy to assimilate, and his background in statistics and probability theory makes his conclusions compelling."

—John Gribbin, author of *In Search of Schrödinger's Cat*

"With this thought-provoking example of a new genre, scientific speculation, Aczel joins the ranks of Roger Penrose, Stephen Pinker, Francis Crick, and others, who combine extensive knowledge of science, a lively imagination, and a flair for stringing words together."

—Keith Devlin, author of *Goodbye Descartes*

"Entertaining, concise, and informative, this book builds its case for extraterrestrial life on solid science. Aczel's range is astonishing: astronomy, organic chemistry, scientific history, mathematics, and the strange world of quantum probability. An invaluable resource, as well as a joy to read. I recommend it highly."

—Nancy Kress, author of *Beggars in Spain*

"Aczel writes with style and grace about a truly fascinating subject. . . . [*Probability 1*] gives a wonderfully clear picture of past and present scientific findings and then in the last few chapters takes a leap into the future through a different method than has so far been widely utilized."

—*St. Louis Post-Dispatch*

"For anyone looking for evidence that we are almost certainly not alone in the universe, *Probability 1* will be an excellent read."

—*Spaceviews*

Contents

Author's Note

SINCE THE ORIGINAL publication of *Probability 1* in 1998, exciting developments have been made in our search for extraterrestrial life. This year, two independent teams of scientists from San Francisco State University and the Harvard-Smithsonian Center for Astrophysics announced their discovery of three large planets around Upsilon Andromedae, a solar-type star 44 light-years away. This discovery marked the first clear evidence of another star similar to our sun, one accompanied by multiple planets in a stable system. Not only did this prove that other solar systems like ours exist, it also heightened the probability of other habitable planets in galaxies far away.

In addition to the discovery of solar systems, numerous advances in the field of science have improved our understanding of how life began on Earth. Chief among these advances is that we know that life on Earth began over 2.7 billion years ago, much longer than previously thought. In relative terms, life on our planet began very shortly after the atmosphere stabilized. It is likely, therefore, that once conditions are right anywhere in the universe, life can begin.

Preface

ON THE EVENING OF February 26, 1998, I was standing on a beach in Aruba with a group of astronomers. Earlier that afternoon, between 2:11 and 2:14 P.M., we had watched the day turn into night as a total solar eclipse passed over the island. We were all still excited from the awesome event we had witnessed only a few hours earlier, and we were now looking at stars and galaxies and a nebula that remained in the sky from a giant supernova explosion witnessed by the Chinese a thousand years ago. As we were talking animatedly about these mysterious objects of the night sky, standing around the large telescope, some of the vacationers at the resort—who had not come here to witness an eclipse or view the stars—came over out of curiosity. After a few moments, one of them moved closer to the telescope and asked Daryl, the astronomer, if she could look through it. He was thrilled to comply and offered to show her and her companion a double star system or a nebula or a beautiful cluster of fifty brilliant red and blue stars called the Jewel Box. "No, no," she said, waving her hand

and smiling. "Could you maybe show us some planets . . . with life on them?"

I walked away down the sand. The quest for life outside Earth seemed more intense now as we approach the third millennium. The woman's wish to see planets with life only served to emphasize this point, which has been made in newspapers and science magazines and television—not to mention announcements from NASA about the findings from the *Galileo* spacecraft, which may have detected evidence for liquid water under the ice cover of Jupiter's moon Europa. Could we finally be on the verge of discovering extraterrestrial life? All this speculation was very exciting for me. I had just finished the manuscript of a book about the probability that life exists on some distant planet in orbit around a star not too different from the Sun—our own star, whose magnificence I now appreciated more than ever, having seen it disappear as by magic behind the Moon and reappear on the other side. Now standing under the bright winter stars, I was thinking about how this adventure started for me.

Just before Labor Day in 1997, I called Jane Isay, executive editor at Harcourt Brace in New York, to ask her if she would be interested in publishing my next book. I had lots of ideas, but Jane liked none of them. So we talked about many other subjects: mathematics and science and probability. And then Jane asked: "How would you like to write a book about the probability of life in outer space?" and she proceeded to tell me how Carl Sagan had wanted to write a book about this topic but for some reason it never materialized, and he passed away. "These are big shoes to step into," I remember saying. "Just try," she answered. I was intrigued.

For a long time, researching this book, I was skeptical. And as I considered the science involved: chemistry and DNA and biology and geology and physics and astronomy, the prospects did not look any more promising.

Then, almost at the last minute, I turned my attention to probability theory. And here, something happened that surprised me. Probability is not an intuitive area. Often, people think they have an answer, but mathematically it does not hold, and something else proves to be true. Mathematics is the key to probability theory, and the math always wins—sometimes despite our intuition. And the mathematics and probability theory always pointed in one direction: the probability of life in outer space is one, just as Carl Sagan had believed. This book will take you through my journey of discovery leading to this conclusion.

MANY PEOPLE HAVE contributed to my research for this book, and it would be difficult to acknowledge all of them here, so I will mention the ones to whom I owe the greatest debt of gratitude. I thank my editor and friend Jane Isay for her vision, encouragement, support, and her trust in me throughout this difficult project. I thank Lorie Stoopack of Harcourt Brace for her superb editing and generous help with the manuscript. I thank Jennifer Mueller of Harcourt Brace. I thank Erin DeWitt for her excellent copyediting. I thank Dr. Michel Mayor of the Geneva Observatory, a great astronomer and a wonderful human being, for his generous help and intriguing conversations. I thank Dr.

Philip Morrison of MIT and Dr. Frank Drake of the University of California at Santa Cruz, founder of the SETI project, for informative interviews. For his views and comments, I thank Robert Naeye, associate editor of *Astronomy*—a magazine I highly recommend to anyone interested in the subject. I thank Professors Marilyn Durkin and Norman Josephy of Bentley College for computer-generated fractals, and astronomer Stephen Mock for various observations. Finally, I thank my wife, Debra, for many suggestions on the manuscript.

1 Fermi's Paradox and Drake's Equation

> There are infinite worlds both like and unlike this world of ours. For the atoms being infinite in number are borne far out into space. For those atoms have not been used up either on one world or on a limited number of worlds, not on all of the worlds which are alike, or on those which are different from these. So that there nowhere exists an obstacle to the infinite number of worlds. We must believe that in all worlds there are living creatures and plants and other things we see in this world.

Thus wrote Epicurus (341–270 B.C.) twenty-three hundred years ago. He had developed some of the ideas about extraterrestrial life put forward by the Greek philosophers Democritus and Leucippus, who lived two centuries earlier. Epicurus put down these thoughts about extraterrestrial life in a letter he wrote to Herodotus.[1]

For the ancient Greek philosophers, the "worlds" referred to here were not planets orbiting stars. The stars were considered part of the firmament of the heaven and were seen to orbit Earth at no greater a distance than those of the planets of our solar system. The "other worlds" were viewed as replications of Earth that could not be seen by observing the sky.

Writing in the first century B.C., the Roman poet Lucretius (ca. 99–55 B.C.) further carried out the Epicurean philosophy of life in the universe. In *On the Nature of the Universe*, he wrote: "Granted, then, that empty space extends without limit in every direction and that seeds innumerable are rushing on countless courses through an unfathomable universe. It is in the highest degree unlikely that this earth and sky is the only one to have been created."[2]

Attacks on these positions were common in antiquity. In his *Timaeus*, Plato (ca. 428–348 B.C.) asserts, "There is and ever will be one only-begotten and created heaven." And Aristotle (384–322 B.C.), the most influential of all Greek philosophers, wrote much about the uniqueness of Earth. It was, in fact, Aristotle's philosophy—which formed the basis for teachings at the universities and for the prevalent religious doctrine in Europe all the way up to the seventeenth century—that prevented the idea of the plurality of worlds from taking stronger hold earlier in history. Aristotle asserted that Earth was the center of the universe and that the Sun, a perfect bright circle in the sky, and the Moon, another unblemished circle, rotated around a stationary Earth together with all the stars in the firmament of the heavens. The theory of the perfection of space and the centrality of Earth to the entire universe was the biggest stumbling block on the road to acceptance of the ideas of Copernicus and Galileo, as well as those of the sixteenth-century Italian philosopher Giordano Bruno, who again suggested a plurality—in fact, an infinity—of worlds.

A century later, Voltaire wrote *Micromegas* (1752), in which he described extraterrestrial life. His story centers on Micromegas, who is 120,000 feet tall and lives on a planet orbiting the star Sirius. Micromegas studies at the Jesuit college of his planet. He derives on his own all of the geometrical theorems of Euclid, and then proceeds to travel to other worlds, including Saturn and Earth. Micromegas has a thousand senses. He complains about life on Saturn, since its inhabitants have only seventy-two senses, which makes communicating with them somewhat less fulfilling than Micromegas would like.

The extraterrestrial life debate reached an apex in the 1850s. In 1853 the British philosopher William Whewell sparked the debate when he published anonymously a book entitled *Of the Plurality of Worlds: An Essay*. The issues raised during the decade-long debate were very much like the advanced and specific elements of the modern argument for life outside Earth. These included the possible existence of planets orbiting stars like our Sun. The fact that variable stars like Algol were observed by astronomers, as well as stars dimmer or brighter than the Sun, led some participants in the international debate of the 1850s to conclude that Earth might be unique and that life might not exist anywhere else. This contention was bolstered by astronomers' inability to observe planets circling any star. These ideas simmered until our own century. The most powerful objection to the theory that life may exist outside Earth came right in the middle of the twentieth century, in the words of one of the most respected scientists of our time.

Fermi's Paradox

In 1950 the nuclear physicist Enrico Fermi dealt a blow to whatever remained of the belief that extraterrestrial life was possible when he asked his colleagues: "Where is everyone?" This question has come to embody what is by now known as Fermi's paradox. By Fermi's logic, if aliens existed, then, since the universe is so old and so large, surely a civilization would have existed that is vastly more advanced than our own and this civilization would have colonized our galaxy. In fact, judging from the experience we've had on Earth, the aliens would have dominated us and taken away Earth's natural resources for their own use, as colonial nations have done for centuries. With a 14-billion-year-old universe and with a galaxy of billions of stars, where are these aliens? Since we haven't seen them, Fermi's paradox suggests, they don't exist.

Some scientists were convinced by Fermi's argument against the chances for the existence of intelligent life in outer space. Most astronomers devoted their research efforts to discovering the physical structure of the universe: how stars form, how they die, how galaxies develop and evolve, and whether the universe shows signs that it might expand forever, as observations from very distant galaxies now seem to imply. It became unfashionable, and in fact dangerous to the prospects for one's career, even to consider the problem of life beyond Earth. The majority of astronomers were not interested in looking for life or even for conditions that might favor life elsewhere in the universe—especially if they did not have permanent jobs or academic tenure. But not everyone felt this way.

Frank Drake was a young doctoral student in astronomy at Harvard University. In 1957 he was working on his dissertation, observing stars through a radio telescope at the Oak Ridge Observatory. Drake had never heard of Fermi's paradox. Had he heard of it, he recently told me, he still would not have paid it any attention. Frank Drake was studying the Seven Sisters.

Tucked behind the head and great horns of the bull of the constellation Taurus is a cluster of stars known as the Pleiades. These stars act as a marker for the Tropic of Cancer, which lies within one degree of their location in the northern skies. Although their number ranges from six to nine, depending on how one counts them, the Pleiades are known as the Seven Sisters. According to Greek mythology, the Seven Sisters were the daughters of Atlas and Pleione: Alcyone, Electra, Maia, Merope, Taygeta, Celaeno, and Sterope. The young girls were pursued by the great hunter, Orion, represented by the next constellation in the sky just west of Taurus. When the gods heard the screams of the distressed sisters, they protected them by placing them as doves in the sky, where they are often depicted as weeping—possibly for the loss of a sister who died, as scientists believe that another star once shone brightly in the region of the Pleiades and has since dimmed. The Pleiades are young stars: they have been forming from a large cloud of dust and gas, whose possible remnants may still be seen through a telescope as a mist enveloping them. The brightest sister is

Alcyone, and the dimmest, hardly visible to the naked eye, is Sterope. Binoculars reveal dozens of new stars forming in this cluster, and a telescope reveals hundreds of them.

As part of his doctoral dissertation project, the twenty-seven-year-old Frank Drake was studying the prevalence of hydrogen in the Pleiades, hoping to learn how new stars are born. "The Pleiades have a very distinct spectrum," Drake explained, "where the hydrogen lines are easily detected." He was looking at these hydrogen lines through his telescope for many weeks in early 1957, when the Seven Sisters were high in the night sky. The patterns of the spectrum were consistent from day to day, with no change, and Drake was performing various calculations that tell astronomers about the chemical compositions of stars. In this case he was trying to find out how much hydrogen is present in these young stars.

One cold night in February, as Drake was looking at the screen of his radio telescope, observing the constant lines from the Seven Sisters, a signal flashed on his screen. This was odd given the usual routine of his observations. Drake straightened up in his swivel chair, his eyes fixed on the interloping signal. A shiver traveled the length of his spine as he understood that the strange signal could not have been caused by a natural phenomenon. Could someone in another civilization in the Pleiades, or beyond, be sending us a signal? Could he be the first person in history—all alone in the middle of nowhere with only his instruments as companions—to receive the call?

A half hour had passed, and the signal was still there on his screen. Drake had a thought. What would happen if he were to shift his radio antenna away from its present direction—would the signal disappear? Very slowly, Drake turned the knob on the control panel of the great radio telescope and heard the motor whir as the dish moved slowly away from the direction of the Seven Sisters. The signal was still there. This, then, was not an alien signal, Drake now knew. He wasn't sure whether to be disappointed or relieved. The signal had to have originated on Earth, or else it would have been direction-sensitive and would have disappeared as the antenna's direction was changed. Frank Drake went home. He needed sleep.

But as he awoke the next morning, Drake realized that the night's experience had left him with something. He could no longer escape the possibility that somewhere out there in space another technologically advanced civilization might be sending us signals. Would we hear their call? As the weeks passed, the notion of being beamed a message by fellow beings in the vastness of space and not hearing their call in the dark obsessed him. Drake became determined that we, as a civilization, must make every effort possible to listen to signals from space. But where should we begin? What was the logical starting point for this cosmic search? Drake kept pondering this question. In the meantime, he finished his dissertation and took his Ph.D. degree from Harvard. He then moved to Green Bank, West Virginia, where the government still maintains large radio telescopes used by a number of groups of astronomers.

The same year, a young professor of astronomy at Cornell University was sitting one day at the concert hall on campus, listening to chamber music. But his mind was not on the music that evening—it was drifting into space. A few years earlier, Philip Morrison had received his Ph.D. degree from the University of California at Berkeley. There, he wrote a dissertation on quantum electrodynamics under the tutelage of the great American physicist Robert Oppenheimer. Listening to the music in 1957, Morrison also came to the conclusion that extraterrestrial beings might well be sending us messages from space, and that we on Earth must make every attempt to listen to such signals. But where should we begin? Both Morrison and Drake, thinking about the same problem independently, reached an identical conclusion: We should scan the microwave band of the electromagnetic spectrum. There, our chances of hearing signals from other civilizations would be the greatest. And we *must* look, no matter how small the odds, both scientists felt strongly. In 1959 Morrison coauthored an article with Giuseppe Cocconi in the prestigious British journal *Nature*. The article would have seemed like science fiction, except that it appeared not long after the Russians launched *Sputnik* and the great superpower competition had moved from Earth into space.

The Cocconi and Morrison article suggested that a civilization on another planet in orbit about a faraway star might have concluded long ago that our Sun was a good candidate for the development of intelligent life on one of its planets. Such an extraterrestrial civilization could be patiently sending us radio signals and expecting that once our technology had reached the stage of being able to detect these signals, we would respond. In reaching their conclusion about which frequency to listen on, Cocconi and Morrison used a clever argument. The most abundant element in the universe is hydrogen, they reasoned, and hydrogen atoms, when excited, broadcast at a frequency of 1,420 megahertz (1,420 cycles per second). This frequency, within the microwave band, is separated from the frequencies where most "noise" of the background cosmic radiation occurs. If an intelligent society were to call us, the two concluded, they would be likely to use this particular frequency.

Green Bank is an isolated valley in a remote region of the Allegheny Mountains. In this area, radio and television stations are few and far between, so that interference from human-made transmissions is minimized. Even cars are few here, so there is less electricity from spark plugs to interfere with the radio telescope's dish. Soon after his arrival at Green Bank, Frank Drake confided in the center's director and managed to convince him to grant him observation time to look for extraterrestrial signals. Drake named the effort he began in 1958 Project Ozma, after the queen of Oz. Drake began by listening to a few radio frequencies from the directions of a number of nearby stars. First, Drake pointed his antenna to the nearby star Tau Ceti (in the constellation Cetus, the Sea Monster). The observations revealed nothing. When this star set behind the horizon, Drake aligned the antenna with another nearby star, Epsilon Eridani (a star lying at about the middle of the River, the constellation Eridanus, west of Orion).

Within seconds Drake heard a strange noise in his earphones—something was pulsating at eight times a second. He felt great excitement, an adrenaline rush. But as a careful scientist, he made himself calm down and study the signal. Then, as suddenly as the signal appeared, it was gone.

He would never hear this sound again. Weeks later it became clear that Project Ozma was not the only one to seek the seclusion of the Allegheny Mountains. Secret military communications experiments were being conducted in these mountains, and these, Drake concluded with no concrete evidence, might have been the source of the interference. Drake turned his attention to other stars. He did not find anything that would bear further scrutiny—all unidentified signals were eventually traced to earthly sources. But the electromagnetic spectrum was very, very wide. Was he looking for signals in the right frequency? Drake could not answer this question. Of all the frequency ranges in the spectrum, the one that seemed to be best was the microwave range. But there were so many others. Drake and his associates reached a conclusion similar to that which Morrison and Cocconi had suggested in their paper. They decided to monitor as many channels as they could within the microwave range, including the hydrogen frequency and those of other elements and molecules they believed were abundant in the universe.

The 1950s and 1960s were exciting times for space exploration. Following both sides' initial successes in space, the United States and the Soviet Union were actively competing for new achievements. In 1961, the U.S. National Academy of Sciences was to hold a meeting at the Green Bank Observatory to discuss the new direction in the study of space: the search for signals from alien civilizations. The sole organizer of the meeting was to be Frank Drake. As soon as he was selected for this duty, Drake started preparing for the meeting. A successful conference, with fruitful, interesting discussions, could result in increased funding for Project Ozma. And the search for alien signals was in dire need of funds: money to pay more astronomers to search the skies, money to develop computing ability, and money to develop radio technology that could search the vastness of the universe more efficiently. Frank Drake felt he needed to come up with a strong agenda for the meeting.

Drake's Equation

Drake's agenda for the 1961 National Academy of Sciences meeting consisted of a single formula, now widely known as Drake's equation. It reads:

$$N = N_* f_p n_e f_l f_i f_c L$$

In this equation, N stands for the number of civilizations in the galaxy currently capable of communicating with other civilizations.

According to Drake, this number depends on the values of the seven factors whose product forms the right-hand side of Drake's equation. What are these mysterious factors that, according to Frank Drake, can tell us the number of advanced civilizations in our galaxy with which we might hope to communicate?

N_*: *The Number of Stars in the Galaxy*

The first term on the right side of the equation is N_*. This stands for the number of stars in our galaxy. According to astronomers, the number of stars in the Milky Way may be as high as 300 billion (although some estimates are lower). This is a huge number. And remember that we are considering only our own galaxy. There are possibly as many as 100 billion other galaxies.

"Oh, Be A Fine Guy/Girl, Kiss Me!"

Let's go beyond Drake's equation. What we need is not just any star, but stars that—from what we know based on our experience on Earth—hold some promise of harboring life. We want to consider stars that are similar to our own Sun. We now know that our Sun is rare: only 5 percent of the stars in our galaxy are Sunlike. Stars on the main sequence (before they collapse and die, turning into white dwarfs or neutron stars or black holes) fall into seven categories of luminosity: O, B, A, F, G, K, and M. (Astronomy students learn the star-type order by memorizing the sentence "Oh, Be A Fine Guy/Girl, Kiss Me!") The Sun is a G-type star, shining yellow, while O and B are hot-burning blue stars; A and F are white, burning at lower temperatures than O and B; and K and M are orange and red, respectively, with less heat and light output than the Sun. The hotter-burning blue and white stars of the O to F categories have relatively short lifetimes, in the millions to a few billion years, while some G-type stars can last even up to 15 billion years or more. The orange and red K and M stars, which can last longer than our Sun, produce too little light and heat. In searching for extraterrestrial life, it is the G-type stars like the Sun that should be of most interest to us. In Drake's equation, therefore, we could multiply the number of stars in the galaxy by 0.05 to count only stars that are similar to our own Sun. This still leaves us many billions of stars.

f_p: *What Is the Percentage of Stars with Planets?*

The next term in Drake's equation is f_p. This is the fraction of stars with planets. When I asked Frank Drake what he thought about the exciting new discoveries of planets orbiting Sunlike stars (the story of which is told in the next chapter), he didn't mince words. "It confirmed in my mind what I already believed: that f_p is equal to about 0.5." This estimate is based on the assumption that the present technology for detecting planets can be vastly improved. Currently, there is an inherent bias toward finding oddball planets: huge gas giants such as Saturn and Jupiter that—contrary to where we expect them to be based on experience in our own solar system—happen to orbit very close to their stars. If we can observe such seemingly unusual planets, what can be said about more "normal" planets, once we are able to detect them as well? The conclusion to be drawn from this seems to be that there are a lot of planets out there!

Michel Mayor, the discoverer of the first extrasolar planet, told me he thought the fraction of stars with planets is 1.0. This means that *every* star has at least one planet. What would make someone make such a strong assertion? Mayor bases his belief on the way planets are thought to form. A star is born when clouds of dust and gas condense and merge because of their mutual gravitational pull. A large lump forms and heats up as it contracts, and nuclear reactions ignite the star; the burning of hydrogen into helium makes the star shine and keeps the matter from further collapse. But as this happens, matter that has not yet condensed remains in a disk that revolves around the star. As the matter in the disk coagulates, objects called planetesimals are formed. These hit against each other, eventually forming planets that continue to revolve around the star in the same direction as the original disk that formed them. If this indeed is the way planets form, there must be plenty of them, since the only way they would not form is if *all* the matter that condensed fell into the forming star—leaving nothing outside for the disk. This is unlikely to happen.

But does life have to evolve only on a planet in orbit around a star? Recent research by Dr. Rudy Schild of the Harvard-Smithsonian Center for Astrophysics indicates that galaxies may contain *quadrillions* of "rogue planets"—planets that are not attached to any star and freely roam the interstellar space within a galaxy. Schild has reached his controversial conclusion using a technique called gravitational microlensing. The principle behind this method is the general relativity concept that light is bent around massive objects. Thus when light from a distant source passes near massive objects on its way to the observer, the light gets focused by the gravitational field of the masses it encounters. When Schild looked through his 1.2-meter telescope at light emanating from a distant quasar, he noticed the microlensing effect around the edges of the distant quasar galaxy. Calculations he then performed with the measurements he obtained led Schild to the belief that the gravitational microlensing was caused by planets. The frequency of the phenomenon then made him conclude that the number of these freely moving planets had to be immense.

From all the information currently available, it is clear that there are at least some planets revolving around other stars. The proportion of such planets, the parameter f_p in Drake's equation, may be high—from 0.5 to possibly close to 1.0. And planets may roam interstellar space as well. Such planets may be captured into orbit by stars they pass, or life may even develop on a roaming planet if it has its own source of internal heating from geological or radioactive sources. We cannot rule out this possibility. Our main concern, though, is with planets orbiting normal stars like the Sun, where life is more likely to develop.

n_e: The Environmental Factor

The next factor in the equation is n_e: the number of planets with environments favorable to the formation of life. This brings us to an important question: What is life? Based on our experience on Earth, we only know one set of conditions that can bring about the development of life. These conditions require abundant water as the solvent through which molecules can travel to form organic compounds leading to proteins. The basic element in these compounds is carbon, to which hydrogen, nitrogen, oxygen, and other elements attach themselves to form the large molecules that are the building blocks of proteins that make DNA and life. Oxygen is essential in the metabolic processes that produce the energy necessary for life. But are these elements the only ones that can lead to life? No one knows the answer to this question. Some scientists contend that life can only exist in the form we know on Earth and that it would require the same elements. Others,

Drake included, believe that it may be possible for life to be based on a central element other than carbon: sulfur, for example. Others think that silicon, at some temperatures, may work similarly to carbon as we know it at Earth's temperatures.

Assuming that water is essential for life brings us to the important concept of the habitable zone. Water is needed as a solvent, and this process only happens when water is in liquid form. So the basic assumption we will use is that planets that can develop and harbor life must have some liquid water on them. Let's look at our own solar system as an example. Here, of nine planets, only Earth is known to have liquid water in abundance. The recent discovery that Jupiter's moon Europa may contain an ocean of liquid water under its icy surface, allowing the possibility that marine life of some kind may have evolved there, may broaden the meaning of the habitable zone. What is striking, however, is that Jupiter and its moons are not usually considered to be within the Sun's habitable zone because, until recently, scientists believed that any water in that region of space had to be frozen solid since the Sun's radiation reaching Jupiter was so weak.

For our solar system, the habitable zone is considered to extend from a point 5 percent closer to the Sun than Earth's orbit and up to a point 37 percent farther away from the Sun than Earth's orbit. Closer to the Sun, photodissociation will occur—the loss of gas and water from the planet into space. The extent of evaporation depends on the planet's mass: a small planet may not be able to hold on to its water even at distances above the lower limit.

This effect apparently took place on Mars. Scientists now believe the planet Mars once had water but doesn't anymore because its small size did not allow it to hold on to its water and atmosphere for much longer than a billion years. The outer edge of the habitable zone is defined as that point beyond which water remains frozen all the time. In the case of our solar system, water remains frozen when a planet is farther away from the Sun by at least 37 percent more than Earth's orbit—but Europa, if it indeed has liquid water, would disprove this theory and show that the habitable zone is not as restrictive as scientists may have thought. The habitable zone for other stars may be defined differently from that of our own solar system. The reason for this is that other stars have different masses and energy outputs, and hence produce different temperatures at various distances.

But n_e in Drake's equation, the environmental factor, requires more of a planet than that it lie within the habitable zone. It also requires the existence of oxygen and organic compounds. Scientists believe that discovering oxygen anywhere in space will be a good indication that life may exist there. Oxygen is very reactive, and it usually is not found on its own but rather as a component of carbon dioxide or other compounds. Thus oxygen molecules or ozone (O_3) are not likely to exist on their own unless they are produced by a metabolic process of life. Recent discoveries of oxygen on Saturn's moon Titan (though in small quantities) give hope for finding some form of life there.

Some scientists believe the fraction of stars with the right environment for life is roughly 10 percent.

f_l: The Fraction of Planets with Life

The term f_l in Drake's equation stands for the actual proportion of planets on which life actually forms. Estimating this term is extremely difficult because we have absolutely no data on planets other than Earth that actually have life. Therefore, any statistically based estimation is impossible. The purpose of this book is to arrive at a *probability* for the existence of life on at least one extrasolar planet. Scientists involved in Drake's search for extraterrestrial life have conjectured that the value of the parameter f_l is about 0.1 or 0.2. It is not clear how they arrived at these numbers.

f_i: Intelligent Life

The term f_i in Drake's equation stands for the proportion of planets on which *intelligent* life evolved. Again, there are no data for any statistical estimates. Later we will discuss the difference between life and intelligent life and review the argument as to whether intelligence is a fluke in the genetic development of life on Earth or whether it is a natural outcome of evolution. When scientists at the 1961 conference discussed Drake's equation, numbers ranging from 0.1 to 0.5 were offered as guesses for the value of this parameter.

f_c: Communication

The term f_c stands for the fraction of planets with civilizations able to communicate with others by radio transmission or other means. Since we have not yet received a single radio signal from outer space, we have no data for any statistical estimation of this parameter. Imagine an intelligent society similar to that of ancient Greece. This is a civilization that has achieved great knowledge, art, political thought, and everything we have come to consider as advanced—except for technology. Such a civilization would satisfy all the other term definitions of Drake's equation, and yet it would have no radio transmitters or receivers. On the other hand, imagine a civilization that is so advanced that it does not require radio-wave transmissions in order to communicate with its members: for example, members of this society may have fiber-optic networks that make radio transmission unnecessary. Part of the idea of the Search for Extraterrestrial Intelligence (the SETI project, the descendant of Project Ozma) is that intelligent extraterrestrial creatures may be detected not only by sending us radio signals intentionally, but also because they may be sending radio signals for their own communications. In both examples above, we have an advanced intelligent society that does not communicate by radio signals and will therefore not be detected by SETI or any other project that listens to radio waves from space.

L: Longevity

Finally, the term *L* in Drake's equation is the longevity of the civilization. The reasoning behind the inclusion of this last factor in the equation is that intelligent civilizations may eventually destroy themselves. Remember that the meeting took place during the height of the Cold War. The assumption on everyone's mind was that intelligent life does not last, since war—nuclear war—will eventually destroy it. This factor, the length of time of a civilization's existence, must be converted to a fraction, by dividing it by the length of time the Milky Way galaxy exists. What was aimed for here was the fraction of civilizations in our galaxy that are still alive, so that we can communicate with them. Of course, because of the tremendous distances involved, there is some uncertainty as to what we mean by this existence and what we mean by communication. If we get a signal from space that was sent fifteen thousand years ago, by the time we receive the signal, the civilization that sent it may no longer exist. A pessimistic estimate of *L* might be from the time of the invention of the radio by Marconi to the time of Hiroshima. This is a very short period of time within the life of the universe. An optimistic estimate might be several million years. When converting *L* into a fraction, we get a small number. The idea is that the window of opportunity for communicating with extraterrestrial civilizations may not be very wide.

AT THE GREEN BANK MEETING, Frank Drake and Philip Morrison met again. Drake had been a science student at Cornell, where Morrison taught. Now they met as the two prominent champions of the search for extraterrestrial intelligence. Morrison's coauthor, Cocconi, did not come to the meeting and, in fact, completely dropped out of the project and eventually became the director of the European Center for Nuclear Research (CERN). The meeting attracted many participants from around the world. Among them was Carl Sagan.

Over the years, Drake's equation received wide acceptance in the scientific community. The ideas it expressed would inspire the search for extraterrestrial life for decades, and the thinking behind this equation would eventually result in the launching of the SETI project (depicted in Sagan's novel *Contact* and in the movie of the same name). The search began on October 12, 1992, exactly five hundred years after Columbus reached the New World, as the giant antenna at Arecibo, Puerto Rico, turned to the sky to look for radio transmissions from other civilizations.

Frank Drake believes that the number of civilizations in our galaxy that can communicate with us is as large as 10,332. He bases this guess on his estimate of the various terms in his equation. The late Carl Sagan believed the number of such civilizations was about 1 million. In this book, we are not concerned with the number of extraterrestrial civilizations in our galaxy able to communicate with us. We are interested in estimating the probability of the existence of life, and of intelligent life, elsewhere in the universe. The first few terms in Drake's equation are very important to us, while some of the others are less so, since these latter conditions may be impossible to estimate in a meaningful, scientific way. Let's concentrate on the crucial terms, the ones we have some knowledge about, and summarize our knowledge about them thus far.

The Goldilocks Search

Once the existence of extrasolar planets is confirmed, the big question that remains is whether any of these planets are suitable for life as we know it to develop. A planet that orbits close to its star is too hot and cannot have liquid water. A planet that orbits far from its star is too cold and its water frozen. Scientists looking for the perfect planet have therefore named their endeavor the Goldilocks search: not too hot and not too cold. Once such a planet is found, other questions implicit in Drake's equation are whether the chemistry is right: Is there water on the planet? Is oxygen present? Are nitrogen and carbon present in the right amounts? These questions can be answered by science. We have methods of analyzing spectra of elements, and we can tell whether any element is present on a planet—if, that is, we can see light from the planet. Once the location of a star with a planet is ascertained, space-borne telescopes could be used to analyze the very dim light from such planets and tell us about the chemical composition of the planet. At present, the expected cost of such a space mission to analyze the spectra of extrasolar planets is about $2 billion, and it is not known whether Congress and the American taxpayers will agree to fund such a project.

Assuming the three basic questions—Are there planets? Are they in the habitable zone? Are their chemistries right for life?—are answered in the affirmative, we are left with the final questions of whether life did indeed evolve and, if it did, whether such life is intelligent. But positive answers to the first three questions will already get us close to our goal. In light of new discoveries and the potential pitfalls of making rash decisions with little or no data, it might be a good idea to take a fresh look at Fermi's putative "paradox."

In pre-Columbian times, a Native American sage could well have said that life on Earth existed only in the Americas. Surely, the logic would go, if life existed anywhere else on Earth, then—assuming there were other continents other than America—life would have developed intelligence, and one of these civilizations would have built large canoes and come here. Since we know of no one who has come here from beyond the ocean, we must be all alone on this "Earth." Well, at some point in time after this "paradox" might have been put forward, someone did come, whether Vikings or Vespucci or Columbus. When people landed on these shores, there was finally contact. The fact that there are time periods prior to the point of contact does not imply any paradox. There must be a *first time* for everything.

It is no surprise that the SETI scientists like the analogy with Columbus. When asked why the SETI project has not yet produced a shred of evidence for any extraterrestrial civilization trying to contact us by radio waves, they point to the voyage of Columbus. "You wouldn't ask Columbus after his first five hundred miles whether he's found a new continent," they say. "We are still on our first five hundred miles."

Unfortunately for the SETI project and its scientists, Congress terminated funding for the search in 1993. But Frank Drake, who heads the project, never gave up. He managed to convince wealthy private donors to contribute to the effort, and their donations now fund the project to the tune of $4 million a year. The sponsors Drake found were William Hewlett and David Packard of the Hewlett-Packard Company, Microsoft cofounder Paul Allen, and Intel cofounder Gordon Moore. Perhaps these donors have some designs for a Cosmic Wide Web.

2 51 Pegasi

EXTRATERRESTRIAL LIFE must have a planet on which to evolve. And the crucial factor in Drake's equation is indeed the fraction of stars that have planets in orbit around them. For thousands of years of human existence, no one knew the answer to the crucial question about the possibility of extraterrestrial life: Are there planets in orbit around other stars? Then a bombshell exploded within the scientific community. It happened in 1995, when two Swiss astronomers discovered the first extrasolar planet.

Late in the evening on the Fourth of July, 1995, two Swiss families from Geneva left their bungalows, climbed up the wooded hill, and entered the dome of the French observatory of Haute-Provence, high in the southern Alps near the village of Saint Michel. They carried baskets of food, bread and cheese, bottles of red wine, and juice for the children, which they spread out on a blanket as one of them pressed the button that opened the observatory's dome, exposing the telescope to the night sky. In a small refrigerator they placed a box containing a large cake and two bottles of fine champagne. It was after 11 P.M. when they finished their meal, and the youngest child in the group was getting tired. "It will happen soon," reassured one of the adults. And they all sat down and waited.

THE MOST STRIKING FACT about astronomy is the immensity of distances and sizes. Space is so vast that the units of measurement used in everyday life lose their meaning. The distance from Earth to the Sun is 93 million miles.[3] A jetliner flying at five hundred miles per hour would get to the Sun if it flew continuously for over twenty years. Contemplating such huge distances forced astronomers to develop a new unit of measurement, one that would fit the scale of our solar system. They defined the distance from Earth to the Sun as one astronomical unit (AU). Mercury, the closest planet to our Sun, orbits it at a distance of 36 million miles, which is just under 0.4 astronomical units. Venus's distance from the Sun is 0.7 AU; Mars's is 1.5 AU; and then come the large outer planets of our solar system: Jupiter at 5.2 AU from the Sun; Saturn, 9.5; Uranus, 19.2, Neptune, 30.1; and Pluto, 39.5. Thus Pluto, the planet farthest from the Sun and the smallest, is 39.5 times farther away from the Sun than we are. A spacecraft can get to Mars within a few months. For it to arrive in the vicinity of one of the outer planets takes several years.

Because the outer planets are so far from the Sun, they get much less sunlight than we do. Jupiter gets 4 percent of the radiation from the Sun that we get on Earth; Saturn, 1 percent, Neptune, .1 percent; and Pluto receives only 0.0006 as much sunlight as we do. Mercury, on the other hand, gets 6.6 times more sunlight than does Earth because it is so much closer to the Sun than we are. Their distances from us and their low luminosity of reflected sunlight make Uranus, Neptune, and especially Pluto, hard to see.

As we look farther out in space, beyond the planets and toward the stars, even the astronomical unit becomes inadequate—using it would be like trying to measure the distances between cities in inches. The stars are so far away from us, and from each other, that to describe their distances we need a much larger unit of measurement: the light-year. One light-year is the distance that light travels in an entire year. Light, which is faster than anything else in the universe, travels at a speed of 186,000 miles per *second*. In one year, therefore, light travels almost 6 *trillion* miles, an incredibly large distance. The closest star to Earth is Alpha Centauri, the brightest star in the constellation Centaurus. The star lies at a distance of 4.3 light-years from Earth.[4] To put the enormous distance between us and this nearest star in perspective, consider the fact that a jetliner flying at five hundred miles per hour will need 8 million years to reach Alpha Centauri. Even our fastest spacecraft, if aided by a gravitational "slingshot" effect of being pulled by the gravity of an approached planet, would require twenty thousand years to reach the star. But in astronomical terms, the stars at Alpha Centauri are our very close neighbors. Our galaxy, the Milky Way, contains 300 *billion* stars. Our solar system lies on one of the "arms" extending from the main part of this spiral galaxy. The diameter of this stupendous collection of stars congregating together in a galaxy spans over 200,000 light-years of space. It takes

50,000 years for the light from the center of the Milky Way to reach us here on Earth. It would be senseless to try to compute how long it might take a spacecraft, which can travel only at a tiny fraction of the speed of light, to get to the center of the galaxy.

The galaxy nearest to our own, the Great Galaxy at Andromeda with its billions of stars, lies 2.2 *million* light-years away. When we view the hazy glow of this galaxy with the naked eye or through a telescope, we are looking 2.2 million years in the past, because it takes this long for the light from the galaxy to reach us.

The known universe contains billions of galaxies. Recently, the Hubble Space Telescope revealed images of galaxies lying at the edge of what we can see in the universe: galaxies that lie at a distance of up to 13 *billion* light-years away. When we look at these very faint images, so incomprehensibly distant from us, we are looking at light that began its journey toward us when the universe was young, 13 billion years ago.

The Sun is much bigger than Earth—109 Earths would fit along the diameter of the Sun, and the volume of the Sun would hold a million Earths. In terms of its mass, the Sun is 300,000 times bigger than Earth. But in stellar terms, our Sun is called a dwarf, reflecting the fact that there are stars so much larger than our Sun that it is dwarfed by comparison. The diameter of Mu Cephei, one of the largest known stars, is over a thousand times larger than that of our Sun, and in terms of its volume, Mu Cephei could hold a billion Suns. Two of the brightest stars in the heavens are also supergiants like Mu Cephei. These are Betelgeuse and Rigel, both in the constellation of Orion. Bright stars such as these burn their nuclear fuel very fast, and consequently they live for relatively short periods of time: several million years. When the dinosaurs looked up to the sky, they did not see these two bright stars in Orion. The stars were born after the dinosaurs had become extinct. Dwarfs like our Sun live for billions of years. Because they are less bright and therefore burn their nuclear fuel at lower temperatures, they are able to live much longer. Stars that are dimmer live even longer—some can live for trillions of years. These stars cannot be seen by the naked eye. The sky, therefore, is full of many more stars than we can see.

For millennia people have gazed at the stars marveling at what these points of light might be. The Babylonians were the first people of antiquity to map the heavens and identify patterns in the stars, which they named as constellations. The Egyptians and the Greeks followed the practice.

A number of twentieth-century astronomers have pointed their telescopes toward the skies in search of planets like Earth orbiting other stars. If life exists outside Earth, it must have a place to develop and harbor it—a planet like our own, orbiting a star like our own to provide it with light and warmth. There have been many false alarms in recent decades when groups of astronomers thought they had detected planets, but these identifications were proven false by other astronomers. The most well-known false identification of planets orbiting other stars is the story of Barnard's Star. In 1916 the American astronomer Edward Emerson Barnard discovered a dim star in the constellation Ophiuchus, which was moving across the sky faster than any other star observed before it. Using photographs of the sky taken during the 1800s, Barnard deduced that his star moved 10.3 arcseconds a year. An arcsecond is equal to 1/3,600 of one degree—it is the thickness of a human hair as observed from ten feet away. In one hundred years, Barnard's Star would move the equivalent of half a full Moon's diameter. This movement indicated that the star might be close to Earth. When parallax measurements were taken—that is, measurements of the angle to the star when Earth was at two separate places in its orbit (say, one during summer, then the other during winter), leading to a calculation of the distance to the star through trigonometry—it was found that Barnard's Star was six light-years away from us. This made the star, a red dwarf, the second closest one to Earth after Alpha Centauri.

In 1938 the astronomer Peter van de Kamp began observing our curious neighbor Barnard's Star from the Sproul Observatory in Pennsylvania. By 1969, van de Kamp had amassed thousands of photographs of the star taken through the telescope, and he observed that the star's motion through space was not perfectly linear: the star was wobbling up and down. Stars move in the sky for a number of reasons. First, the stars in the Milky Way circle the center of the galaxy as it rotates through space. Then there is the motion of all the stars (detectable in observations of distant galaxies) away from each other due to the expansion of our universe. This phenomenon was discovered by Edwin Hubble in 1925, and the rate at which the universe is expanding is known as Hubble's constant.[5] Star motions, leaving out the nightly motion due to Earth's rotation, imply a straight-line trajectory. The wobbling in the observed motion of Barnard's Star told van de Kamp that a planet—and later he decided that there were even two of them—orbited the star, and that the gravitational pull of these planets was causing the star to wobble.

Until as late as 1930, even our own solar system was not known completely to astronomers, and another planet was yet to be discovered orbiting the Sun. Photographing the night sky through a telescope, tiny patch after tiny patch and then photographing the same areas of the sky at a later time and making a comparison by flipping the pictures one after the other led to a great discovery in the early part of this century. In 1929 the Lowell Observatory in Flagstaff, Arizona, acquired a new photographic telescope and hired Clyde William Tombaugh, a twenty-three-year-old farmer from Kansas with a passion for astronomy, to search the skies for another planet orbiting our Sun. At that time, only eight planets were known, Uranus and Neptune having been discovered in the nineteenth century. The observatory's founder, Percival Lowell, had one objective in the latter part of his life: to find a new planet orbiting the Sun, the chimerical Planet X. Tombaugh had built a nine-inch reflector telescope on his father's farm, which he constructed from parts of farm machinery and the crankshaft from his father's 1910 Buick. He drew pictures of the planets Mars and Jupiter as he observed them from his telescope and sent these drawings to Mr. Lowell. Finally, the invitation came from Lowell to work at his observatory. Tombaugh did not believe in mathematical computations others had told him would reveal the hiding place of the mysterious planet. If a planet was to be found, he felt, it would be due to his own thoroughness in compiling observations. He worked continuously for a year, tirelessly taking thousands of pictures of the sky and comparing them by flipping. On February 18, 1930, he discovered Pluto near a star in the constellation Gemini. Tombaugh became the most celebrated astronomer of his time. He was the last person to find a planet circling our Sun—no other "Planet X" was to be discovered, even though some astronomers continued to believe in the existence of yet another large planet orbiting the Sun. (Since Pluto is smaller than our Moon, some astronomers still consider it a large asteroid rather than a true planet.)

Tombaugh succeeded in his search because even though Pluto is very dimly visible through a large telescope—it orbits the Sun at a distance of roughly 40 AU from us—its very weak light could still be seen. As we go beyond our solar system, however, distances are measured in light-years (trillions of miles, not hundreds of millions), and detecting the light of a planet—which is reflected from its star—would be an impossible task. What adds to the problem is that the light from our Sun, for example, is a *billion* times brighter than the reflected light from even the largest planet, Jupiter. Discerning this very weak light as distinct from the light of a star a planet orbits would be a daunting task. In looking for extrasolar planets, therefore, astronomers began to look for tiny wobbles in the proper motion of a star, which could then be attributed to the gravitational pull of an unseen planet in orbit around the star. Such expected movements would be very, very small because the mass of even a large planet like Jupiter is typically a thousand times smaller than that of the star it orbits. A very powerful telescope would be required to detect the wavy motion.

But in 1969 Peter van de Kamp believed he had detected precisely this kind of wobbly movement in the path of Barnard's Star. Could it be that the first extrasolar planet had been discovered? In 1971 George Gatewood, then a graduate student at the Allegheny Observatory, began to study Barnard's Star with the observatory's thirty-inch telescope. Within a few months, Gatewood was able to show that the deviations away from a straight line that van de Kamp had observed in the movement of Barnard's Star were simply due to disturbances caused by Earth's atmosphere. There was no planet orbiting the star. It became clear that if astronomers were to detect the very slight wobbles within a star's motion that could be attributed to the pull of an unseen planet, they would need more powerful telescopes.

WITH A DIAMETER OF 1.9 meters, the telescope of Haute-Provence is hardly among the largest or most powerful in the world. The observatory was built sixty years ago, at an altitude of 650 meters above sea level—not an altitude where powerful observatories are built. But the observatory was equipped with a highly sophisticated system developed by the two Swiss astronomers, Michel Mayor and Didier Queloz, who were now waiting patiently at the telescope with their families. The two astronomers were based at the Geneva Observatory but were buying observing time from the French at Haute-Provence because the less powerful observatory was available for longer periods of time than observatories with more powerful telescopes, where astronomers had to compete for single nights of observation. At the telescope's eyepiece, Mayor and Queloz had attached a spectrograph—a device based on a diffraction grating with attached electronic components—which separates starlight into its wavelengths. Wires connected the device to a small computer. The two astronomers were specialists in measuring the Doppler shift in the spectral lines of stars.

The Doppler effect is a familiar phenomenon from everyday life. If you stand at a train station and a train goes by blowing its whistle, the sound will appear to change pitch from high to low as the train passes you. An approaching train has a high-pitched whistle sound, and a receding one a low-pitched sound. In the nineteenth century, the Austrian physicist Christian Doppler (1803–1853) first studied this effect of sound waves, now named for him.

Light waves and other electromagnetic radiation also exhibit the Doppler effect. In 1905 Albert Einstein presented his special theory of relativity. His results were contrary to the common intuition of scientists of his time, who thought that light—as everything else in the universe—should change its speed when its source is moving. For example, if a football player runs with the ball at 10 MPH and then throws the ball as he is running, the speed of the ball with respect to the ground will be its speed if the player were stationary *plus* 10 MPH. In the early part of this century, two physicists, Albert Michelson and Edward Morley, tried to find evidence that light, too, behaved in this way. They believed that light moved through space by means of an unseen "Ether," and they were looking for changes in the speed of light as it moved through the

Ether. Since Earth revolved around its axis, they reasoned, the Ether had to drift with it. By beaming light from one mountain in California to another, in a line that coincided with Earth's direction of rotation, the scientists expected to find that the speed of light was increased by the speed of Earth's rotation in the same way that the ball's speed was increased by the player's running speed. But they found no such change, no matter how hard they tried and how extensive were the resources they expended on this project. At one point, Michelson, frustrated with his inability to detect the Ether drift, made an elaborate proposal to the state of California for what he said would be a project to survey the mountains in the Sierra Nevada range. But the officials knew what he was after and turned down his request. It was then that Einstein's purely theoretical work with pencil and paper and no measurements at all proved that the Ether drift couldn't exist. The speed of light, roughly 186,000 miles per second (300,000 kilometers a second), was a universal absolute, Einstein said. And this speed did not change, no matter how fast the source of light was moving with respect to a stationary observer. With Earth's rotation or without it, the speed of light remained the same.

But while the speed of light does not change as the source of light moves, the *frequency* of the light waves, and the associated wavelength, do change. And these are shifted precisely according to the Doppler effect. When a source of light moves closer to the observer, the frequency of the light waves increases. It's as if the waves were "crowded together" by the forward speed of the source. Green light, for example, will be pushed to higher frequency and will appear to a stationary observer as blue, since blue light has a higher frequency (and a shorter wavelength) than green. Conversely, when the source of light is moving away from the observer, its wave function is "stretched" by the speed away from the observer, and the frequency will shift downward. A green light may thus shift to red when the source moves away at some speed. This effect on light is called a redshift. If you drive your car at a speed of 90,000 miles per second (half the speed of light) and approach a red traffic light at an intersection, the light would appear to you as green. But don't try this argument in court—you'll get a speeding ticket instead.

Doppler studies have been used by astronomers for the past two decades. When a star is moving away from Earth, its light will exhibit a redshift. When a star is moving toward Earth, its light will shift in frequencies toward blue, violet, and ultraviolet wavelengths (depending on how fast it is moving with respect to the observer). To detect the actual amount of the shift in frequency, and thus to be able to measure the star's proper motion (motion relative to an observer on Earth), scientists need to decompose the star's light waves into their components. These elements of the star's spectrum are called its spectral lines, and they can be revealed by a spectrograph.

In the 1980s various teams of astronomers around the world started to look for Doppler shifts in the spectral lines of stars to try to detect motions caused by the gravitational pull of unseen planets. These teams were very competitive, since the first team to observe spectral shifts that could be attributed to a planet orbiting a star other than our Sun was certain to receive worldwide acclaim. The main teams involved in the search for extrasolar planets were Bruce

Campbell and Gordon Walker of British Columbia, Canada; a team in San Francisco, Geoffrey Marcy and Paul Butler; and the Texas team of Artie Hatzes and William Cochran. The Swiss team of Mayor and Queloz (Mayor's graduate student at the University of Geneva) entered the race very late—in April 1994, seven years after the California team had begun its work and thirteen years after the Campbell-Walker team had started on its search for planets.

The theory behind using Doppler analysis to look for planets that may orbit other stars is an interesting one. The largest planet in our solar system is Jupiter, with a mass 318 times greater than that of Earth. Even with such a huge mass, Jupiter is still only 0.1 percent as massive as our Sun. Yet, as Jupiter revolves around the Sun, it exerts a small gravitational pull on the Sun. This effect is due to the fact that every action causes a reaction. Jupiter moves in a nearly spherical orbit around the Sun because the Sun exerts a gravitational pull on the planet. Without the gravitational pull of the Sun, Jupiter would fly off into interstellar space in a straight line. But the reaction to this force is that the Sun, too, exhibits some motion due to the pull of Jupiter. What really happens is that *both* Jupiter and the Sun perform a circular dance around their common center of mass. Since the Sun is so massive compared with Jupiter, the common center of mass lies much, much closer to the Sun than to Jupiter. In fact, the center of mass of this two-object system lies inside the Sun, but away from its center in the direction of Jupiter. Thus as Jupiter moves around our Sun, the Sun circles the center of mass of both of them.

Suppose an observer were to stand at a distant location in space in line of sight of the planets' orbits. Looking at the Sun with a highly sensitive spectrograph, this observer would detect Doppler shifts in the spectral lines of the light from the Sun as the Sun moved toward and away from the point of observation. For as Jupiter rotated away from the observer to circle the Sun, the Sun would move toward the faraway observer at some speed as it was rotating around the Sun-Jupiter center of mass. And as Jupiter was appearing behind the Sun and moving toward the faraway observer, the Sun would move around the center of mass back and away from the observer. These cyclical motions of the Sun toward and away from the observer would produce slight Doppler shifts in the Sun's spectral lines in the way that the pitch of a train's whistle changes as the train approaches the observer and then recedes.[6]

The astronomers specializing in Doppler shifts were well aware of this phenomenon. Thus, they reasoned, if planets orbited other stars like our Sun and if we happened to view these stars in the correct direction—that is, in a line running through the plane in which such planets revolve around their stars—a slight Doppler shift in the star's spectrum would reveal the existence of the star's motion in response to planets' gravitational pull. This Doppler shift would provide evidence for the existence of such planets. The expected effect was very, very small, measured in meters per second. If it could be detected, however, calculations could then reveal a minimum mass for the planets (minimum because the angle of observation may not be perfectly straight on) and their distances from their stars. The trick was to devise a method that would reveal such tiny changes, and then to be able to perform the complicated calculations quickly. The Swiss latecomers to this game knew how to perform both tricks. Over months at Haute-Provence, Mayor and Queloz had analyzed the spectra of 142 stars. These stars were all relatively nearby—at distances of less than sixty light-years away from us—and all of them were of the same type as our Sun. By October 1994 they had still found nothing.

To measure the Doppler shift correctly, astronomers need a reference point so that they can measure exactly how much the lines have moved. Various compounds emit light at particular wavelengths (producing spectral lines) when they are heated. If you take a diffraction grating and look at the yellow sodium streetlights through the grating, you will see the characteristic yellow spectral line of the sodium element in the street lamps. The green mercury lights used in some places have their own spectral lines when viewed through a diffraction grating. The Canadian team devised a calibration instrument containing the highly toxic compound hydrogen fluoride, which has very distinct spectral lines, making the calibration very accurate. The San Francisco team of Geoffrey Marcy and Paul Butler used the much less dangerous element iodine for their calibrations. Marcy and Butler looked at the spectra of many stars through the telescope of the Lick Observatory in the Santa Cruz mountains of California.

But Marcy and Butler were bogged down by computational complexity. While their instruments were more accurate and more powerful than those of the Swiss team, enabling them to detect stellar motion as slow as three meters per second in comparison with the Swiss team's twelve meters per second, the computer program that Marcy and Butler used to analyze their data was horrendous. So instead of stopping their observation activity from time to time to do the necessary computations and find out whether they had found anything, Marcy and Butler simply stored their data on the hard disk of their personal computer, with the intention of sitting down someday to do the calculations when their observations were finished. But unknown to them, and to the rest of the world, evidence for the existence of an extrasolar planet was already hidden inside the hard drive of their computer.

Mayor and Queloz did not get involved with toxic or noxious chemicals. To calibrate their spectrometer, the Swiss team used a simple tungsten lamp. When heat from electricity excited the electrons in the filament, the electrons moved to higher energy levels. Once they descended back to lower levels, the electrons emitted light of specific frequencies as predicted by the quantum theory. Mayor and Queloz used these spectral lines to calibrate their spectrometer. Their accuracy was twelve meters per second, enough to detect a roughly Jupiter-sized planet. And the computing algorithm that the Swiss astronomers had devised was vastly superior to those of the other teams. This enabled the two scientists to compute the Doppler shifts at the observatory while their telescope was still aimed at the star they were studying. All 142 stars the team had selected were G- (yellow) or K- (orange) type stars, similar to our Sun in size and luminosity. In September 1994 Michel Mayor and Didier Queloz aimed the Haute-Provence telescope at a star of medium brightness located just below the Great Square of the constellation Pegasus. This Sunlike star fifty light-years from Earth was named 51 Pegasi.

A favorite of poets and painters, Pegasus is the winged horse of mythology. He was conceived when the sea god Poseidon disguised himself as a horse and seduced the Gorgon Medusa. When Perseus killed Medusa, Pegasus sprang from her body: a fully formed horse with wings. Pegasus was tamed by Bellerophon and Athena and ascended Mount Olympus, where he carried Zeus's thunderbolts. The Great Square, visible at its highest level in the sky during September, comprises four stars—Markab ("saddle" in Arabic), Algenib ("side"), Scheat ("shin"), and Sirrah ("navel")—and makes up the main body of the horse.[7]

As soon as they trained their telescope on 51 Pegasi, lying below the Markab-Scheat side of the Great Square, Mayor and Queloz noticed that this star was different from the rest of the stars they had observed. Something did not look normal. After obtaining several measurements of the spectrum of the star, Michel Mayor decided that the star was so strange that they should reobserve it for a longer period of time. In December 1994 Mayor and Queloz spent an entire week observing 51 Pegasi. Of the 142 stars on their list, this one still made no sense: the speed obtained from the Doppler calculations was sixty meters per second—this star was racing around in a circle! Mayor's first reaction was that something had gone wrong with the spectrograph. How else could he explain that after over a decade of searching for radial star motion by a number of astronomers worldwide nothing had been found, but within a few months of trying he and his colleague discovered a star moving radially at such an incredible speed?

The two astronomers went back to look at other stars, and there they found no such effects—the spectral lines did not shift in a cyclical way. It wasn't the spectrograph, Mayor concluded—it was 51 Pegasi. In January 1995 they returned to observe the star. Slowly, their calculations revealed a pattern: something big was circling 51 Pegasi, but not as big as a brown dwarf. Brown dwarfs are failed stars, which cannot support life. Stars are born when dust and gas gravitate together over thousands of years. Slowly, the matter forms what astronomers call a protoplanetary disk. As matter is mutually attracted to other matter, a motion develops in the disk and it starts rotating around itself. Eventually, matter forms planetesimals: objects the size of small rocks revolving together. Thousands of years later still, the center of the disk enlarges as more and more matter succumbs to the gravitational pull of the central objects.

As the matter condenses due to the strengthening gravitational pull of the unifying object, it heats up and starts to glow. If enough matter has accumulated so that the object is at least 10 percent as massive as our Sun, eventually the gravitational forces produce such high pressures and resulting high temperatures that the object ignites nuclear reactions in its core: converting hydrogen into helium by nuclear fusion. The object becomes a star. The tremendous heat and radiation that result from the nuclear reactions prevent the matter in the star from further collapsing inward. The star then remains stable and radiates light and heat as well as other kinds of rays for millions to billions of years. If the object is not massive enough to ignite its nuclear reactions, it does not become a star. It is then considered a brown dwarf. Brown dwarfs release little light, if any, and are thus not observable.

Many stars in our galaxy, the Milky Way, are double-star systems, and some are even triple stars. These multiple stars jointly circle their center of mass. Over 60 percent of the observable stars in the Milky Way have been shown to belong to such systems. In the last few years, some star systems have been demonstrated to consist of a star and a brown dwarf rotating together. Astronomers define brown dwarfs as objects greater in size than eighty Jupiters. The first thing that Mayor and Queloz tried to do was to ascertain that the object they seemed to have discovered was not a brown dwarf, for there would be no chance of life on a failed star and therefore no glory in the discovery.

As a star forms, the remaining material in the disk around it continues to rotate. Millennia later, the matter in the disk keeps encountering the collective gravitational forces of the rest of the matter and slowly begins to fuse together. According to the standard theory in astronomy, this is how planets form around a Sunlike star. Astronomers based their theory solely on what they had observed within our own planetary system. Our solar system has large gas giants in outer orbits far from the Sun: Jupiter, Saturn, Uranus, and Neptune. It has small rocky planets: Earth, Venus, and Mars at inner orbits close to the Sun. Therefore, astronomers generalized these observations to conclude that any other planetary system that might exist in the universe must follow this pattern. The thinking was that heavy elements such as iron and rocks rotated close to the center of the disk, while lighter elements such as hydrogen and helium rotated in the outer parts of the disk. Hence, small and rocky planets must form close to the newborn star, while large amounts of gas and water in the form of ice, with some rocky material in the center, formed the distant large planets like Jupiter.

It was this theory that kept Mayor and Queloz puzzled for so many months. The object they believed was circling the star 51 Pegasi was not a brown dwarf, but it was very massive. They were able to compute its mass tentatively as at least 60 percent the mass of Jupiter. But when they looked further at their numbers, their calculations didn't make sense. This "planet," if it was one, was orbiting its star at a distance of one-eighth the distance of Mercury's orbit around our Sun! In fact, the object causing the Doppler shift in 51 Pegasi's spectrum had to be only 5 percent as distant from the star as Earth is from the Sun. What was a planet almost the size of Jupiter doing so close to its star? This finding contradicted the whole accepted theory of planetary formation. As they were again checking their instruments and their computations in March 1995 and were nearing a conclusion with a few more necessary observations, the Winged Horse and its stars were about to disappear from northern hemisphere skies for a period of four months. The constellation would not return until July.

What made these astronomers in the United States, Canada, and Europe so convinced that planets could be found around other stars that they would devote decades of hard work to the effort to discover them? The answer to this question is that they had received a tremendous boost to their hopes in 1991. In that year, not one but three extrasolar planets were discovered. But the reason these planets were not celebrated in the world media was that the star the planets were orbiting was dead.

Once a star is formed and begins to radiate light and heat, its ultimate fate depends solely on its mass. A star with a mass up to somewhat less than eight times the mass of our Sun will keep burning hydrogen into helium and helium into heavier elements in nuclear reactions lasting several billion years. Once it has expended all of its nuclear fuels, the star will become a red giant as its size increases, and consequently it will shed its outer layer, which will float away in what astronomers call a planetary nebula. The core of the star will then condense and become a white dwarf, remaining in this stage indefinitely.

If, however, a star begins its life with a mass of over eight times that of our Sun, its fate will be very different. Such stars burn bright—much brighter than our Sun. Because of the tremendous output of energy from these stars, the stars burn themselves out very quickly, in astronomical terms. A large star will expend its nuclear fuels within millions of years, rather than billions of years as is true for our Sun and the stars in its class. Because of the high temperature and radiation intensity released from large stars, they appear blue—blue light being of higher energy levels than other colors of light. When the star's life begins to end as its nuclear fuels run out, the star will increase in size to become a red supergiant. Then the star explodes. The explosion is called a supernova, and its light can be seen as an exceptionally bright light in the sky—brighter than the most luminous star. In A.D. 1054 Chinese astronomers recorded a bright light in the night sky that lasted several months. Today, we know the event recorded by the Chinese was a supernova. The remnants of that explosion can still be seen in the sky as the famous Crab Nebula (which astronomers call M1) in the constellation Taurus.

Once the supernova occurs, much of the star is gone—blown off into space at tremendous speeds. The remnant of the explosion, still a very massive body, then collapses inwardly with little resistance. As the gravitational forces push matter together, the matter changes. Atoms are no longer able to exist as such. Electrons and protons, forced together by tremendous gravitational forces, collapse to form neutrons. Matter as we know it everywhere in the universe no longer exists inside the dead star. If the collapse stops there, the star, now made almost exclusively of neutrons, is called a neutron star. Typically, the diameter of the star will be about ten or more miles, but its mass will be much larger than that of our Sun. A neutron star is thus a fantastically dense object. A spoonful of its matter weighs as much as a mountain on Earth. If the original star is even more massive, the collapse continues beyond the neutron star stage. The diameter of the dead star will continue to decrease as the star collapses and its density will approach infinity. Eventually, what remains is a black hole: an object of close to zero diameter with close to infinite mass density. Now the gravitational pull of the object is so tremendous that not even light can escape it, hence the name black hole.

A neutron star will often rotate. When a neutron star rotates, it does so at an amazing speed: several complete revolutions per second. Because of the fierce magnetic field produced by a rotating neutron star, it emits powerful radiation directed into space as a beacon. A rotating neutron star, emitting radiation at regular intervals so precise that clocks can be set to them with better accuracy than atomic clocks on Earth, is called a pulsar. Pulsars were first discovered by Jocelyn Bell, a graduate student in radio astronomy at Cambridge University in 1967. Bell discovered a radio source that was transmitting a burst of radiation every 1.3 seconds from a location in the constellation Vulpecula. At first she thought she had discovered a transmission from an alien civilization. But the lack of a Doppler shift in the transmissions indicated that the source could not be on a planet. Since that time, astronomers have discovered over eight hundred pulsars. The pulsar that lives inside the remnant of the A.D. 1054 supernova observed by the Chinese is one of the youngest known, and it spins once every 0.03 seconds. Some of the pulsars are microsecond pulsars: neutron stars that spin so fast that they emit bursts of radio waves milliseconds apart. These are believed to be very old neutron stars that have stopped their radio bursts as even this energy source within them had been exhausted. The stars are now truly dead. But then, at some point, their tremendous gravitational force pulls into them some interstellar matter from a nearby star or other object. As the matter hits the surface of the neutron star, the star begins to spin again, but this spin is so fast that it takes milliseconds to complete a single revolution. The result is a millisecond pulsar.

In 1983 a young Polish astronomer by the name of Alex Wolszczan moved to Puerto Rico, where the Arecibo telescope, the largest radio telescope in the world, is located. The telescope was malfunctioning, and while being repaired over a period of years, its thousand-foot antenna could not be aimed in the direction of the galactic plane as was usually done when radio astronomers were using it. This fact gave Wolszczan a unique opportunity to use the powers of the great telescope to look for pulsars in unpopular directions in space, which were of no interest to anyone else. No one competed with Wolszczan for observing time, and he could have one of the most expensive pieces of hardware used in astronomy all to himself.

By late 1990 radio astronomers had detected a number of millisecond pulsars, and Wolszczan's research using the Arecibo dish revealed unexpectedly a new millisecond pulsar, later to be named PSR B1257+12, whose beams were weakly detected from the direction of the constellation Virgo. The source was determined to be located at a distance of 1,300 light-years from Earth and to spin once every 6.2 milliseconds—or over 161 times *per second*. Having fortuitously found a source located in exactly the direction at which the crippled giant antenna was aimed, Wolszczan could take his time studying his pulsar at leisure. He enlisted the help of Dale Frail, a radio astronomer working at the Very Large Array (VLA) of twenty-seven radio telescopes in New Mexico. Using two different locations to study a single source of radio beams in the sky allowed for great accuracy and the ability to pinpoint the location of the source, in the same way that seeing an object with two eyes allows us to estimate its distance from us much better than if the object is seen through one eye only.

By December 1991 Wolszczan and Frail had detected slight perturbations in the signal from PSR B1257+12. Since pulsars send radio waves with such precise regularity, they act like cosmic clocks of unmatched accuracy. Thus any slight shift in the radio wave can be detected with no ambiguity. The two radio astronomers implemented a computer program that analyzed the transmission shifts they were observing. Within weeks they were able to establish a curious fact: PSR B1257+12 had *three* different planets orbiting it. Using a complex equation based on Kepler's laws, they determined that these planets were roughly the size of Earth and were orbiting the dead star at distances close to those of Earth and Venus from the Sun. While the results that Wolszczan and Frail had obtained are unquestioned due to the fantastic accuracy of pulsar emissions, the mystery remained: What were planets doing in orbit around a star that had exploded millions of years earlier in one of nature's most cataclysmic events—a supernova? If planets had orbited around the original giant star, the explosion would have blown them into space. The only conclusion scientists could support was that the planets had formed around the pulsar after the supernova had destroyed the original star. Since the star was dead and emitting no light or heat, the chances of life on these planets were very slim. Furthermore, a pulsar's radiation is so intense that any life that might have developed on the planets would have been summarily destroyed by the lethal bursts of gamma rays and other high-energy radiation. The fact remained, however, that in 1991 three planets were discovered orbiting a center other than our Sun. This finding encouraged all who were on the lookout for planets that might harbor life: planets around normal stars like our Sun.

In October 1995 an international conference was to be held in the Italian city of Florence, where a number of

Doppler-shift astronomers were to present papers. Mayor and Queloz were anxious to finish their observations—they feared that some of their competitors might have already found planets orbiting normal stars. They were eager for Pegasus to return to the night sky so they could complete their research in time to present their findings in Florence.

Just before midnight on July 4, 1995, the Winged Horse's head became fully visible in the eastern sky, and with it the Great Square. Just off the Great Square, the two astronomers could see 51 Pegasi with the naked eye. They pointed their telescope and made the measurements they had been waiting to make for four months in order to have a complete set of readings of the movements of the star over a number of cycles. But for greater accuracy, they needed the star to transit, that is, pass as high overhead as possible. This would only happen at 4:00 A.M., so a little after midnight, Mayor and Queloz sent their families to bed. If all went well just before dawn, they would celebrate together in the morning. Four hours of waiting had passed, and Mayor and Queloz conducted their final test. They could now prove definitively that a planet of at least 60 percent the size of Jupiter orbited close to the star 51 Pegasi, revolving around it every 4.2 days. Back with their families in the early morning, the astronomers popped open the champagne and celebrated. Everyone enjoyed the raspberry cake, including the youngest child, Didier Queloz's three-year-old son, still too young to understand the reason for the great excitement.

MAYOR AND QUELOZ wrote their conclusions in a scientific paper and readied it for presentation at the Florence conference in October. Within days of the announcement of the findings by Mayor and Queloz at Florence, Geoffrey Marcy and Paul Butler verified their results independently. The planet orbiting 51 Pegasi became fact. Within the next year, Marcy and Butler analyzed their own observations and reported the existence of a number of other extrasolar planets orbiting stars similar to our Sun. Other teams discovered more planets. By 1998 nine extrasolar planets orbiting Sun-like stars had been discovered. (A tenth planet was observed by the Hubble Space Telescope, but it seemed to have been ejected from its solar system in the constellation Taurus.) Could some of these planets harbor intelligent life? Of the nine extrasolar planets, one—the planet orbiting the star 70 Virginis—was believed by scientists to orbit within the star's habitable zone. Does life exist on that planet?

3 The Chemistry of the Universe

IN 1864 A STRANGE METEORITE that looked like burned matter was brought to the laboratory of the great French chemist and microbiologist Louis Pasteur (1822–1895)—the man who gave his name to the pasteurizing process he invented. The unusual object had fallen to the ground at 5:30 P.M. on March 15, 1806, in a loud explosion that was heard in the air above the village of Valence in southern France. Two farmers working in a nearby field saw a large object plunge to the ground nearby and rushed to inspect it. What they found was like nothing they had ever seen before. The object resembled a large rock, but it was made of a softer material and was all black—it looked like a piece of burned rubble. The curious item was taken away to be analyzed by scientists, who discovered to their astonishment that it contained 20 percent water and 10 percent *organic* matter. This finding was surprising given that the object was believed to be a meteorite.

Meteorites are objects that fall to Earth from outer

space. Usually, they are categorized as either irons, which are meteorites made of iron and nickel, or stones, which are pieces of extraterrestrial rock. The stones are by far the most common type of meteorite, but they are hard to find because they bear a strong resemblance to common Earth rocks. The irons are easy to detect on the ground because they are denser than rocks and when inspected carefully they reveal their metallic composition. But the meteorite found at Valence was very different.

In his *Memoir on the Organized Bodies Which Exist in the Atmosphere,* Pasteur had shown that the air contains germs and that when these germs are isolated in a medium where they can grow, as in a petri dish, they multiply and can be identified. Pasteur invented sterilizing techniques that he used to separate organisms and determine where and when they exist, and he constructed a special drill to remove samples from the center of the carbon-based meteorite brought to him. Using this technique, Pasteur hoped to determine whether life had existed at one time inside the meteorite, before any possible contamination from Earth. Extracting uncontaminated samples was especially important since the meteorite had been touched by humans and had been kept on display in museums before it was brought to his laboratory. Pasteur removed small amounts of matter from the meteorite using sterile techniques. He then inserted the samples into a medium where any organic matter would then grow and develop. Pasteur found no organisms in the meteorite, although there was no doubt that it contained the complex carbon compounds we call organic.

Today scientists know that about 5 percent of all meteorites that fall to earth every year are of the kind found at Valence. Such meteorites are called carbonaceous chondrites because they contain carbon and organic compounds. They are believed to have formed in space when our solar system was young: they have been dated to 4.5 billion years ago. These remnants from the formation of our solar system arrive on Earth in the same way as do the other meteorites, the stones and the irons. New and rigorous chemical analyses of the French meteorite showed high-weight molecules: paraffins, tar, fatty acids, and other carbon-based molecules. These are extremely important compounds in the search for extraterrestrial life, since these are the very same carbon-based compounds that give rise to protein and other building blocks of living organisms.

Another carbonaceous chondrite, the Murchison Meteorite, landed in Australia in 1969. Analysis of this meteorite brought a big surprise: this object contained fifty amino acids—eight of which form the basis for all proteins. Carbonaceous chondrites give us direct evidence that organic material exists in outer space. But what is organic material? Is it life? And what are the chemical elements and processes that take place in space?

The stars in the heavens are factories of chemical elements. If the big bang theory is correct, the universe began with a gigantic explosion about 14 billion years ago. In the explosion, huge amounts of matter and energy—which Einstein taught us are really equivalent to one another—were blown into the space they created and began to expand along with it. At first, radiation and lightweight particles were all that existed, and as these traveled at enormous speeds outward and collided with one another, baryonic matter was created from these smaller particles. Baryonic matter is matter as we know it: electrons, protons, neutrons, and atoms of elements.

The lightest element in the universe is hydrogen. A hydrogen atom contains only two particles, with opposite charge: a positively charged proton and an electron with an equal but negative charge. The nucleus of a hydrogen atom consists of a single proton, and the electron revolves in a spherical orbit around the proton. As electrons flying out after the big bang met protons, the electric attractions brought them together to form a tremendous number of hydrogen atoms. When two such atoms met, they combined in a chemical reaction to form the hydrogen molecule, H_2. Scientists believe that the most common constituent of the early universe was hydrogen (along with various isotopes of helium and lithium).

The primeval cloud of hydrogen spread through creation and condensed in large patches that continued their outward travel. Over time, clouds of hydrogen condensed more and more tightly because of the mutual gravitational force that matter exerts on other matter around it. And as a cloud of hydrogen became more and more dense, it heated up. Eventually, the heat and pressure were so great that nuclear reactions were ignited: the hydrogen atoms began fusing to form helium and in the process released large amounts of energy. Two hydrogen atoms were no longer comfortable to stay in the form of a hydrogen molecule—two atoms bound together by electrical charge forces. The heat and pressure made the two nuclei fuse together and join two neutrons to form a nucleus, with electrons now orbiting the combined helium nucleus that was formed. The energy that was released by the reaction was in the form of heat and radiation, including light (these are the reactions that occur inside our Sun). Thus a second element, helium, was born in a universe containing mostly hydrogen.

In 1868 Sir Joseph Norman Lockyer (1836–1920) discovered helium. But he did not find it on Earth. Lockyer deduced the existence of the element helium inside the Sun by observing spectral lines in sunlight that had not been identified before. Lockyer thought that the element he discovered was a metal and hence gave it the name helium: the "hel" part from the Greek word *helios* for Sun, and the "ium," as in sod*ium*, calc*ium*, and so on—the ending signifying a metal. Ironically, the helium we find on Earth did not originate in the large reservoirs of helium in the Sun. Any helium that came to Earth when the planet was formed has since dissipated into space. The small amounts of helium we do find on Earth originate as products of the decay of some radioactive elements in the ground.

Once a star burns its hydrogen into helium, it faces gravitational collapse and death unless it can burn further. The star might then start to burn the helium in its core to form a new element: carbon. The birth of carbon was the most fortuitous event in the history of the universe—for this is what allowed life to develop on Earth. Without the miracle of carbon, life as we know it could not exist. Why is carbon so important? The answer lies in its chemistry.

Unlike nuclear reactions, which occur in the nucleus of an atom, chemical reactions link one atom with another through the exchange or sharing of their electrons. A carbon atom can form electron-sharing chemical bonds with as many as four other atoms. These can be other carbon atoms, or atoms of other elements, or a combination. An amazing variety of possible combinations of atoms can form carbon-based molecules, making carbon the most versatile element in the universe.

A diamond, for example, is a single giant molecule made only of carbon atoms. Each atom is strongly bonded to four other carbon atoms, in the corners of a tetrahedron. It is this crystalline structure that gives the diamond its hardness as well as its aesthetic qualities. But the same carbon atom can bond to just two oxygen atoms, sharing two electrons with each, forming the gas carbon dioxide, CO_2. Another molecule of carbon is methane, which consists of carbon bonded to four hydrogen atoms spatially distributed around it: CH_4. Methane and carbon dioxide exist in abundance where carbon is found on planets. Earth's early atmosphere consisted mostly of these two gases. The emergence of life on Earth changed this composition. Carbon also likes to form long chains or rings with other elements—hydrogen and a few other types of atoms. Thus carbon is literally the backbone of many large and complicated molecules we find in nature.

Here we enter the vast realm of organic chemistry. The name derives from the fact that many such compounds come from living things. We know millions of different organic compounds—a tribute to carbon's incredible ability to bond in a seemingly endless variety of ways. The "organic" compounds that came from outer space in the form of the carbonaceous chondrites were exactly such large molecules of carbon atoms linked together, with hydrogen and other elements around them. So open space, too, seems to contain the miracle atoms of carbon. But where did all this carbon come from?

The young universe consisted of simple gases—hydrogen and helium. Once stars burned in nuclear reactions converting hydrogen to helium and then helium to carbon, they could go on for a while and sustain themselves against gravitational collapse. Then, stars could burn still further, creating other elements: nitrogen, oxygen, phosphorus, and finally iron. Once iron was formed, the nuclear reactions must subside, since fusion involving iron would take away energy, rather than produce it. When a star transformed all its matter into iron, it would die. The death would occur in the form of a supernova, which creates elements heavier than iron, or the formation of a planetary nebula.[8] The planetary nebulae enrich the universe with lighter elements, including carbon. Once enough stars had lived out their lives of millions to billions of years, the universe might have enough elemental matter floating freely through space so that matter could settle in significant amounts to form a chemically rich protoplanetary disk. This is why our solar system was formed "only" 5 billion years ago, while the universe as a whole began somewhere around 15 billion years in the past. Objects that were created earlier than our solar system may not have contained the chemical wealth necessary for the development of life. For example, it is unlikely that life as we know it would develop on a gas giant somewhere—a planet like Jupiter made of hydrogen gas and an ice core.

Some scientists have contended that life could only develop around a star belonging to a large galaxy such as our

Milky Way, because only large galaxies have enough matter in them to form the necessary numbers of stars, which—once they die—could enrich the galactic space with chemical elements in the abundance necessary for life to develop. The iron that made the steel in your car—and the iron in the hemoglobin in your blood—once resided deep inside a large star that exploded in a supernova in the distant past. The carbon forming the long chains that make up your genetic code was once spewed into space as a star died long ago. The past history of the universe from the big bang billions of years ago is what brought us to where we are today. In a sense, we are all incarnations of dead heavenly bodies from the cosmic dawn.

The chemical "soup" that permeates some parts of space where stars have died and their contents spilled to fill cubic light-years of interstellar void becomes reactive as chemical elements mix. Using spectral analysis of light that passes through the interstellar clouds, scientists have been able to determine the existence of complex molecules. The organic molecules in the carbonaceous chondrites thus formed in deep space as hydrogen atoms met carbon. But carbon doesn't always wait for stars to die before it goes traveling in space. Sometimes it launches itself.

Scientists were stymied by the existence of a few variable stars that did not have a regular pattern of brightening and dimming that could be attributed to an unseen companion eclipsing the star. The foremost among these irregular variable stars discovered was R Coronae Borealis. It is a member of the constellation Corona Borealis—the Northern Crown, a beautiful arc in the northern skies. Every few weeks R Cor Bor, as this star is known informally, dims by a number of orders of magnitude. Scientists could not figure out why this happened, until they used spectral methods to see what is in the star—what lines in the spectrum are absorbed when the star dims. The results were surprising: the star gives off huge clouds of soot—carbon particles. This carbon covers the environment around the star for a period of time, masking the star's light. When the carbon clouds drift into space, the star appears bright again—until the next flare-up of soot. But where does all the carbon dust go? Space is empty, isn't it?

We know from observations of deep space that space is not uniform. Light-years around the Earth and Sun, space is very empty, with about one particle in a cubic centimeter of space. This is what allows us to send spacecraft into Earth orbit, or to the Moon, or to other planets and beyond, as we do.

But there are large areas of distant space where matter is much denser. The Eagle Nebula, eight thousand light-years away from us, is a region stretching over several light-years in every direction, with gargantuan clouds of dust and gas containing a wealth of elements. And the same is true for other large areas of space. Nothing could travel through such dense clouds of matter. In them, matter gets denser and denser as it coagulates to form stars and planets. Conceivably, space around us used to look like this 5 billion years ago. Eventually, the huge clouds condensed to form the Sun and the planets. Possibly, the nearest stars, the system of Alpha Centauri four and a quarter light-years away, were also made from the same cloud—if the cloud stretched that far when our solar system came to be.

Galaxies are full of these star-forming clouds. Telescopic observations of faraway galaxies reveal many nebulous areas. In the Milky Way, even a small amateur telescope can show us star-forming areas. The Orion Nebula, called M42, lies at the bottom of the sword hanging from the hunter Orion's belt. This is a cloud with an area of *fifteen light-years* in diameter, lying fifteen hundred light-years from Earth. The region is visible with the naked eye as a nebulous spot in the night sky. Within it, stars are forming by the thousands from the large cloud.

But the most amazing cloud of all is the one that lies in the direction of Sagittarius. In the summertime in the northern hemisphere, look south just above the horizon and you will see the constellation of Sagittarius, the Archer. The constellation is associated with a satyr—a creature that is half human and half goat, with a horse's tail. He holds a bow and arrow in his hands, aiming the arrow at the center of the giant scorpion in the sky, the constellation Scorpius. Sagittarius, however, doesn't look as much like an archer as it does a teapot, and that is what makes the constellation easy to identify. Look for a tilted teapot, an old-fashioned one, handle on the left, a triangular top, and the spout on your right. You can easily imagine boiling water pouring down into a cup. When you look in the direction of the teapot's spout, you are looking in the direction of the very center of our Milky Way galaxy. Here, a curious fact emerges.

Our galaxy contains hundreds of *billions* of stars. So, when we look in the direction of the center of the galaxy, lying thousands of light-years from us, we should see many more stars than we see in any other direction in the sky.

And yet, this is not the case. We see Sagittarius, and Scorpius to its right, but nothing more—the sky is not full of bright stars in this region as we expect. The reason for this apparent paradox became clear recently when the Hubble Space Telescope was aimed in the direction of Sagittarius and began observing space in the infrared frequencies. Suddenly, the sky in this direction became very bright, in fact, much brighter than scientists had expected. It was found that in the direction of the center of the Milky Way lie huge stars that are millions of times brighter than our Sun. The reason we can't see these stars in the visible-light range is that the direction of the center of the galaxy is covered with a tremendously large and thick cloud of gas and dust. Here, millions of stars are forming, but we can't see a thing. So our galaxy contains huge amounts of "dark matter," as scientists call the invisible clouds of dust. The huge unseen quantities of matter flowing freely inside the galaxy provide it with the chemical richness that gave us life. Such clouds gave us the abundance of carbon that is the central element for the creation of life, along with hydrogen and other elements. What are these other elements?

Oxygen: The Source of Energy for Life

Oxygen is the most common element in Earth's mantle. By weight, oxygen constitutes 89 percent of water, 23 percent of the air (21 percent by volume), and it makes up 50 percent of the weight of the common silicate minerals. This abundance of oxygen is again due to the death of the stars that created oxygen in their nuclear reactions. Oxygen is higher in molecular weight than carbon, so the fusion reactions that create it inside stars are of a higher level. The oxygen atom will combine with other atoms in chemical bonds that involve the sharing of two electrons. This is half the number of connections that carbon can make (four) and results in a smaller number of possible bonds.

Oxygen, however, is an extremely reactive element. It is this reactivity that makes life possible. When oxygen combines with other elements, energy is released; and this energy, when utilized within a living organism, sustains life. We've seen a common chemical bond involving oxygen: the carbon dioxide molecule, CO_2. Here each oxygen atom shares two electrons with the carbon atom. Water is another common molecule of oxygen: H_2O. These are strong bonds of oxygen, and both carbon dioxide and water are very stable molecules. When there are no other atoms around, oxygen atoms will combine in pairs: sharing a pair of electrons with each other. But like an unhappy couple, each partner too similar to the other, the bond breaks when other, more attractive elements are around. This is exactly what makes oxygen so reactive. When a carbon-containing compound comes in contact with oxygen and initial energy is provided by a spark or by heat, the matter will burn: the oxygen atoms will free themselves of each other and both will combine with the carbon atom to form the strongly bound CO_2 molecule.

The hydrocarbon rings and long chains get attacked by oxygen when the conditions are right: a spark inside your engine when oxygen and vaporized gasoline come together will cause the reaction that moves the car. The combustion reaction produces the two stable molecules—carbon dioxide and water. When a carbon-containing substance is burned in the presence of not enough oxygen, the less stable carbon monoxide molecule, CO, is produced. Here the bond is not optimal since the carbon wants to share more electrons. Hence CO is an unstable molecule that will react with oxygen-bearing hemoglobin in the blood, eventually poisoning a living creature that breathes it in high concentration.

The opposite process, that of the release of oxygen from carbon dioxide, is what's hard to carry out and requires living organisms to accomplish. Astronomers believe that other planets will have little free oxygen in the form of molecular O_2 or the less stable ozone molecule, O_3, unless these planets have life. The reasoning is that there is no way for the stable water and carbon dioxide molecules to release their oxygen unless a living plant metabolizes the carbon dioxide. As plant life developed on Earth, the photosynthesis reaction using the plants' and trees' chlorophyll began to fill Earth's atmosphere with free oxygen. After many millions of years of this process—which uses sunlight to break the CO_2 molecule, releasing the oxygen and providing the flora with energy—the atmosphere began to contain the large amounts of oxygen we now see (21 percent of the air volume), allowing the development and sustainment of animal life. When we burn fossil fuels, we tip the scale of oxygen to carbon dioxide in our atmosphere away from oxygen and toward carbon dioxide. When we cut down and burn trees, we are committing a double crime, since not only are we burning away our oxygen and producing the greenhouse gas CO_2, but we are also removing from Earth its wonderful trees, which provide us with fresh oxygen.

Because of the lack of oxygen in large quantities except where photosynthesis releases it from CO_2, the search for planets with free oxygen is a search for extraterrestrial life. Recent scientific observations from Saturn's moon Triton suggest the existence of some amounts of ozone, and earlier observations of Jupiter's moons Europa and Ganymede also indicated the existence of at least some oxygen. If this can be confirmed by observations from space in the next few years, there may be hope for finding life on some of these satellites of planets in our solar system. Two Sunlike stars—Epsilon Eridani, at a distance of 10.7 light-years from Earth, and Tau Ceti, at 11.8 light-years away—have spectra indicating carbon and oxygen. This means that these two elements are present in each star's atmosphere. However, we don't know whether these stars have planets, and if they do, whether oxygen and carbon are present on them. Free oxygen in the atmosphere of a planet around a nearby star similar to our Sun would constitute wonderful news for anyone interested in the search for extraterrestrial life.

Nitrogen: The Basis for Proteins

All proteins are nitrogenous substances. They contain 16 percent nitrogen, along with carbon, hydrogen, oxygen, and sometimes other elements such as sulfur, phosphorus, iron, and copper. Without the crucial element nitrogen, there would be no life as we know it on Earth.

Nitrogen likes to bond with three hydrogen atoms, sharing an electron with each to form ammonia, NH_3. This pungent gas occurs where organic matter decomposes, as in manure piles. The ammonia molecule is more stable than the complicated proteins, and thus it is produced in the decay. Ammonia exists on some of the planets in our solar system, along with water ice and methane. These compounds, containing the elements carbon, hydrogen, oxygen, and nitrogen, are believed to be present elsewhere in the universe. Two nitrogen atoms form a stable triple bond between them, sharing three electron pairs and making the stable N_2 molecule. Eighty percent of the volume of the air in the world consists of nitrogen in this molecular form.

Sulfur: Does It Offer an Alternative to Carbon?

Sulfur is a heavier element. Does it hold any promise of occupying the place of carbon as the backbone of life processes elsewhere in the universe? When sulfur burns, we get sulfur dioxide, although sometimes we get sulfur oxide or sulfur trioxide. The variety of possible bonds of sulfur is an indication of the element's versatility, and indeed, sulfur likes to form chains and rings that are somewhat similar to those of carbon. A commonly occurring compound is a ring of eight sulfur atoms, S_8. Sulfur will also form long chains of sulfur atoms, one linked with the other.

Frank Drake told me that he thought extraterrestrial life could possibly be sulfur-based, and I believe he came to that hypothesis because of sulfur's apparent ability to form large molecules. But sulfur, while able to form long chains or large rings, has limited potential. The reason is that sulfur does not possess the richness of possible bonds with other elements. Carbon forms long chains or rings to which *other* elements—hydrogen, oxygen, nitrogen, and others—are bound. Sulfur's long molecules tend to include only sulfur. This property of sulfur does not allow for the wide spectrum of compounds required for the development of life.

What about Silicon?

As an alternative to carbon, silicon has also been discussed as a possible base element for life. The reasoning here is a little more logical, from a chemical point of view, since silicon has the same valence as carbon (4). One might expect silicon to work similarly to carbon when it combines with itself and other elements. But this does not happen. Silicon (along with boron and germanium) is a *metalloid*—its chemical properties are between those of metals and nonmetals. Metals, regardless of their chemical valence, do not form compounds that are similar to the life-bringing carbon bonds. Being in between a metal and a nonmetal, silicon forms the basis for quartz (hexagonal crystalline silicon dioxide), glass (boron and aluminum silicates), and portland cement (calcium silicates), as well as sand, mortar, and asbestos. It is hard to visualize life arising from any of these compounds.

Interestingly, coesite—a compound containing silicon tetraoxide—was discovered in Meteor Crater in Arizona. From this and other finds, we know that silicon exists in outer space, but it has been observed in rocky things, as it has on Earth. When silicon bonds with hydrogen and oxygen, it cannot form long chains or rings with other silicon atoms, as carbon does with itself. The longest such molecules contain three silicon atoms in succession, clearly not enough to give us hope of finding extraterrestrial life based on compounds of silicon. It seems that the best bet for such life will be on compounds based on carbon. From all the evidence, carbon is uniquely qualified to be the basis for all life-related compounds. This will become clearer in the next chapter, on the molecules of life.

The Role of Metals

The hemoglobin in our blood contains iron, and without iron we would not get oxygen to our body's cells and couldn't live. Crabs have another molecule in their blood for delivering oxygen, based on copper. Our nerves work on a principle of the electric potential of sodium and potassium ions, and our bones and teeth consist of compounds of the metal calcium. In addition, we require zinc and a number of other metals in small amounts in order to maintain good health. Metals, which are generally heavy elements, are produced in the supernova explosions of massive stars, and we are fortunate to have these essential metals to sustain life as we know it.

Crystals

Most solids are crystalline in nature. A crystal consists of atoms arranged in a repeating three-dimensional pattern. The regular arrangement of the atoms in the crystal gives the crystal its characteristics, in particular, the shape of a polyhedron—a figure bounded by plane faces, such as a cube or a tetrahedron. We've encountered a crystal earlier:

the carbon crystal, a diamond. The structure of crystals—the configuration of the atoms or molecules in the crystals—is studied by the diffraction of X rays. This science was developed by the German physicist Max von Laue (1879–1960), who discovered the phenomenon of the diffraction of X rays by the atoms in the crystal, and by the British physicists W. H. Bragg and W. L. Bragg.

IN THE EARLY 1970S, while an undergraduate student at the University of California at Berkeley, I was working at the Lawrence Berkeley Laboratory as a research assistant to my chemistry professor. Gabor Somorjai is a Hungarian chemist who escaped to the West during the 1956 uprising against the communists and found his way to Berkeley. Somorjai is an expert on surface chemistry: the study of chemical reactions that take place on the surface of a crystal. Within a few years, Somorjai had managed to build an excellent research group at Berkeley. He had, at any one time, seven or eight doctoral students who spent all their time bombarding crystals of platinum with low-energy electron beams. Each experiment was carried out inside a steel chamber with almost perfect vacuum. The walls of the chamber were plated with gold to prevent any unwanted chemical reactions. An electron beam was created electrically within the chamber and aimed at the platinum crystal in the center of the chamber. Each of Somorjai's students had his or her own system, and I knew each system cost—in those days—over a hundred thousand dollars.

Gabor Somorjai had one of the largest research budgets of anyone at Berkeley. The money came mostly from oil companies, since Somorjai's group was developing the prototype for the catalytic converter now used in cars. A catalyst is something that affects the rate of a chemical reaction. The catalytic converter is a catalyst that promotes—that is, speeds up the rate of—conversion of polluting carbon products in the car's exhaust system to less polluting compounds. Chemical reactions within the human body are promoted or inhibited by catalysts as well: these catalysts are called hormones. At any rate, Somorjai's students experimented with the structure of platinum since their professor believed that the chemical reactions required to convert the bad carbon products into less polluting ones might take place on the surface of a platinum crystal.

Platinum is not reactive. Like gold, it will only react with a very strong acid called aqua regia ("royal water," a mixture of nitric and hydrochloric acids) and with fused alkalies. So it would seem unlikely that compounds of carbon that result from the combustion of hydrocarbons should be involved in reactions with this precious metal. But Somorjai understood well how surface chemistry works. He knew what he was doing.

My job was to grow the costly platinum crystals. Then I would use a spark cutter, an electric blade that cut the crystal by producing electric sparks, cutting through the metal like butter. This technique was used to protect the crystal structure from the destruction that would result from usual cutting, which applies force to the crystal. Finally, I would perform X-ray diffraction of each crystal to ascertain the crystal's orientation, the direction of the atoms on the face of the crystal: horizontal, diagonal, or at an angle to the lattice. The X-ray picture of the face of the crystal will have a pattern of spots on it, and each regular pattern of spots told a different story about how the lattice of atoms was oriented.

For many months Somorjai's team wasn't getting anywhere. The chemical compounds were not reacting on the surfaces of the platinum crystals, as the researchers would find from the electron beams they were shooting. The crystals I was cutting to the researchers' specifications were all very "ordered": the atoms sat in regular rows on perfect planes facing up or diagonally in one of various ways. Then one day I got a special request. "I need a stepped surface," Somorjai told me. "Cut the crystal somewhere *between* the 100 face and the 111 face." This was an unusual orientation. Instead of looking at a plane of atoms head-on or looking at it in the direction of diagonal rows, the professor wanted something in between. He explained that when this was done, the resulting crystal would form microscopic steps—it would break along the surface so that "corners" would form with one direction head-on and the other diagonal. At each "corner," Somorjai hypothesized, the atoms will be reactive since they will not be bound to many neighbors. Within days of the experiment, Somorjai's group became even more prosperous—the idea of the catalytic converter was discovered in the professor's laboratory, and the car manufacturers were ready to buy it for the device the government wanted them to install in all cars.

One late evening not long after the breakthrough, Somorjai called me into his office. He knew I had other interests and would soon be leaving his lab. "I am not interested in cars or catalytic converters," he said. "You see, I want to discover the secret of life." I looked at him surprised. I'd thought his entire research enterprise was geared toward making money from the automotive industry. He smiled, sensing my puzzlement. "The molecules that make up living things are very, very complicated," he said. "We are dealing with long, involved chains and rings connected in extremely intricate ways. Usual chemical reactions will not produce them: putting chemicals in a test tube and then shaking it. . . . You have to have a catalyst. I believe that, eons ago, life started on a stepped surface of a platinum crystal."

4 The Double Helix of Life

WHAT IS IT THAT distinguishes a living organism from an inanimate object? In his book *General Chemistry*, the late great American chemist Linus Pauling (1901–1994) defines the chemical reactions that characterize life.[9] Pauling points out that a plant or an animal has the power of reproduction, which a rock—whatever chemical reactions may affect it—does not. A living thing has the ability to produce progeny sufficiently similar to itself to be recognized as belonging to the same species. Pauling continues by noting that living things ingest food, subjecting it to chemical reactions involving the release of energy and the secretion of products of these reactions. This process is called metabolism. And living things also have the ability to respond to their environment. A plant may direct itself toward the Sun, and animals move in search of food or a potential mate. Pauling points out some difficulties in the definition of life. Plant viruses may replicate themselves by using the host plant's genetic material but do not have the ability to ingest foods or move on their own. This is an exception to the general definition of life. A virus, however, even if not considered a complete living organism because it lacks the powers of locomotion and metabolism, is still extremely complex: it is a molecule with molecular weight in the order of 10 million.

Living things consist of cells, and the more complex the organism, the larger the number of specialized kinds of cells it contains. An amoeba consists of a single cell, a simple plant may have only a handful of types of cells, and a human being has many kinds: blood cells, nerve cells, cells that secrete hormones and acid in the stomach, liver cells, lung cells, muscle cells. The main contents of a cell are water and proteins. Proteins are very large molecules, with weights ranging from 10,000 to many millions (the atomic weight of carbon, for example, is 12, so the smallest protein molecule contains the equivalent of roughly a thousand carbon atoms, although in reality it also contains hydrogen, nitrogen, oxygen, phosphorus, sulfur, and copper or iron). A red blood cell, for example, is 60 percent water, 5 percent various materials, and 35 percent the iron-containing protein hemoglobin, with molecular weight of 68,000. The human body contains many different kinds of proteins, each of which carries out a specific life-supporting function. Proteins are composed of certain organic acids called amino acids.

Organic compounds—that is, substances that are based on carbon atoms linked with each other and with other elements—can sometimes come in two forms: one the mirror image of the other. The best way to visualize this is to imagine two gloves. No matter how you look at it—toss it in the air, rotate it, turn it upside down—a right glove will never turn into a left glove. And yet the two are really the same item: both gloves have a central part for the palm, four fingers, and a thumb. The difference between the two gloves is simply their orientation. Depending on which side of a central part of an organic molecule is linked to a side configuration of atoms, it can be a right-handed molecule (denoted D, for the dextro- form), or its mirror image, the left-handed form (L, for levo-) of the same molecule. Every amino acid except for glycine comes in these two forms. And here an extraordinary biological phenomenon emerges: plant and animal proteins contain exclusively the L-form of all amino acids. No one has been able to explain this curious fact. If chemical compounds are obtained by random chemical reactions, one would expect statistically that half the amino acids in living organisms would be of the L-form and half of the D-form. Chemically, the two forms of each acid are identical, and yet we find only one of them in nature, everywhere.

Around the turn of the twentieth century, the German chemist Emil Fischer (1852–1919) studied proteins and discovered that they contain long chains of amino acids, called polypeptide chains. The formation process by which amino acids are added on can continue, one amino acid after the other, to result in a long chain. Chemists have developed methods for determining the number of polypeptide chains that make up a protein molecule. The hemoglobin molecule, for example, contains four polypeptide chains of amino acids. But chemists did not know how to determine the three-dimensional shape of the polypeptide chains of amino acids in a protein, and this arrangement of the atoms in space was very important for understanding the chemical properties of protein. The breakthrough came in the early 1950s by the use of X-ray diffraction methods. Linus Pauling was the first to determine the actual structure of a polypeptide chain. The structure he discovered was a helix: a spiraling coil of carbon, nitrogen, hydrogen, and oxygen atoms. Some of the coils he found were spiraling upward in a clockwise fashion and some in an anticlockwise direction (thus some were right helices and some left helices), but all of the amino acids on the chains were of the L-form (left-handed) as everywhere in nature. This aspect of the behavior of proteins remained an enigma.

But how do we get from amino acids and proteins to life? In 1922 the Russian biologist Aleksandr Oparin presented to the Russian Botanical Society a paper with his theory about the origin of life. Oparin imagined life beginning in the oceans, as chemical elements dissolved in the water mixed together and, energized by sunlight, formed the molecules essential to life. In 1955 two scientists at the University of Chicago, Stanley Miller and Harold Urey, tried to create life in a bottle using Oparin's idea. They filled a container with what they thought might have been Earth's primeval makeup: ammonia, methane, water, and hydrogen. Then they simulated lightning by making electric sparks pass through the mixture. Within a few days, they found in the mixture some amino acids—the building blocks of protein. But they didn't create *life* in the laboratory—and neither has anyone else since then.

In the fall of 1951, James D. Watson and Francis Crick found the structure of the genetic molecule of life, deoxyribonucleic acid (DNA). They used X-ray diffraction photographs of crystals of DNA. This molecule was so complex, it was the only one capable of replicating itself and thus allowing a living thing to grow and to produce offspring. The DNA molecule held the secret of life. Watson was impressed with Pauling's discovery of the helix of protein and came to believe that DNA—though a much more complicated molecule than protein—might have a similar helical structure. A helix packs a lot of matter into a relatively small three-dimensional space, so a larger molecule was even more likely to use this efficient structure.

In inorganic chemistry, one works with molecules that have no ambiguity of structure. A water molecule, H_2O, has only one possible structure—the two hydrogen atoms linked with the oxygen atom with only one possible angle between each hydrogen and the oxygen atom. In organic chemistry, there are many possible ways for the carbons to be linked with each other in a chain or a ring, and there are many possible ways for the side elements or submolecules to be tied to the carbon backbone of the large molecule. So determining how an organic molecule is held together in space is crucial for understanding what it is and how it behaves. Watson started to put together a helical model out of the metal parts produced in the laboratory's shop. He thought a helix had to be the right model, despite the fact that new X-ray pictures were inconclusive about a helix.

Science knew some things about DNA. The molecule's nature was somehow beyond that of proteins or sugars or hydrocarbons or anything else: this molecule had vast amounts of information coded right inside it. But how? What was the coding mechanism? And how would the code be read to produce an entire animal or plant or a human being? From chemical analysis, scientists knew that the DNA molecule contained more than its sugar-phosphorus-nitrogen base submolecules that formed the backbone of the huge collection of atoms. Somewhere in the molecule were four kinds of very special substances. These were two pyrimidines, six-membered rings made of carbon and nitrogen, to which hydrogen atoms were attached. The two compounds were thymine (T), and cytosine (C). Two other compounds, called purines, were also attached within the DNA molecule. These were compounds consisting of two attached rings of carbon and nitrogen, nine atoms in twin cycles with four hydrogens bonded to some of them. The two molecules were adenine (A) and guanine (G). Somehow, these four molecules—A, G, T, and C—arranged in a variety of combinations, formed the language by which the DNA molecule communicated the genetic code of the individual. But how? And where were these "letters" attached to the backbone of the DNA structure?

The pattern that emerged after much work was a double helix. It looked like a licorice twist. On each side was a basic chain of molecules, sugar, phosphate, and a nitrogen base—the two strands wrapped around each other in a double helix. The genetic code molecules A, G, T, and C were attached between the two long backbone molecules. As soon as the model was finished, Crick and Watson stopped to look at it. There was something beautiful about the array of atoms in front of them. The arrangement was too elegant *not* to be the true structure of DNA. In 1962 Francis Crick and James Watson received the Nobel Prize for the discovery of the structure of DNA.

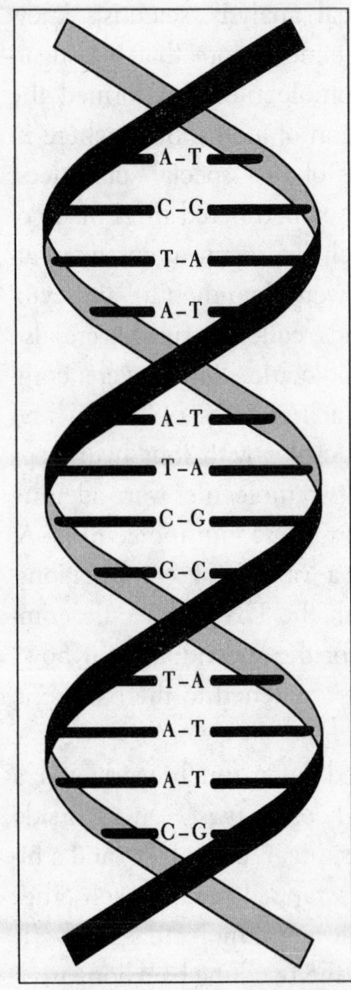

What is that elegant megamolecule whose structure Crick and Watson had discovered? What is this spiral staircase with steps labeled A, C, T, and G, whose arrangements determine the structure and function and generation of life? The two outside chains of the large molecule are wrapped around each other in a helix. But inside the helix, the four basic genetic information holders—A, T, G, and C—are not randomly placed. Adenine, A, only links to thymine, T, on the other string of the helix. The two submolecules are held each to its string, and they "join hands" with each other in two places: A and T form two hydrogen bonds with each other. A hydrogen atom belonging to adenine is attracted and is linked to an oxygen sitting across from it on the thymine molecule. The same bond occurs between a hydrogen atom belonging to thymine, which then links up with a nitrogen atom on the adenine molecule. Cytosine and guanine only link with each other, and they do so using three hydrogen bonds: one from C to an oxygen atom on G, and two from G to a nitrogen and an oxygen atom on C.

If we look at one side of the double helix only and identify the molecules on it, we will find a succession of letters, for example: A, A, G, C, T, G, G, C, C, C, T, A, and so on. We are now reading information. We are reading the genetic code of a living organism. All living things contain the DNA molecule. This, in itself, is a striking fact. Here is a chemical compound whose purpose is to encode vast amounts of information in an amazingly compact and efficient way. To understand just how vast the information is that's coded on a DNA molecule, consider that everything about you as a person, from the color of your eyes and hair to the shape of your knees and the length of your fingers and toes and everything else in between, is coded on a molecule—a tiny object that cannot be seen by the naked eye, and whose length is measured in angstrom units (10^{-8} centimeter each). The elements A, T, G, and C lie along the length of the double helix separated from each other a distance of 3.3 angstroms. The double helix is 20 angstroms in diameter. How is the information coded? What *is* the code?

The DNA double helices reside inside genes. These genes live inside chromosomes, which in turn occupy their place inside the nucleus, which is at the heart of the living cell. In the DNA molecule, the sequence of bases—A, G, C, and T—is coded into sequences of three-letter words. Thus, A-G-A or C-T-G, or G-G-T might be such words that encode information. With four possible letters, arranged into three-letter words, we have a total of $4 \times 4 \times 4 = 64$ words in the DNA lexicon. The words in the DNA sentences tell the cell the exact sequence of amino acid residues (proteins) to be synthesized by the cell controlled by the particular gene.

Today, we can identify the specific genes that can cause inherited disease and advise potential parents about the risks to their future children's health. Our genes are now known to determine in large part our height, weight, appearance, and intelligence. Studies of identical twins adopted by different parents and raised without contact with each other have revealed stunning, unexpected similarities in everything from character, appearance, likes and dislikes, intelligence, leadership ability, creativity, and even the choice of a career and hobbies and other interests. The DNA in our genes controls our life far beyond what we might expect.

When DNA structure was finally discovered, the question arose as to how the molecule effects the passing of its attributes from one generation to the next. The Watson-Crick mechanism by which the DNA molecule replicates itself works as follows: First, the double helix starts to uncoil itself like a snake. Then, the bonds holding the two chains together through the bases inside start unlocking themselves. At the same time, a new chain is synthesizing right along the uncoiling DNA molecule and develops its own bases attached to the chain. The same happens with yet another new chain. Finally, bases from the new chain attach themselves to the old DNA chain in exactly the allowed format: A attaches itself to T and vice versa; G attaches itself to C and vice versa. Thus the composition of the mother DNA molecule is preserved in the two daughter molecules by the laws of chemical bonding that dictate A-T and G-C combinations only.

Life seems to be controlled and organized by a complex system governed by a giant, intricate compound called DNA. This DNA appears in two autonomous levels within each of our cells—the nucleus and the mitochondria—telling each cell exactly what to do and when to do it. It tells the cell how to metabolize sugar into energy, how to dispose of waste products, how and when to split to form new daughter cells. The molecule itself contains a tremendous amount of information coded inside it. This information determines all the characteristics of the species—no matter how advanced or complex it may be—and each attribute of the individual within the species, down to the minutest detail. And when the individual of the species mates, the DNA in its genes determines how the information about the individual will be passed on to the new generation.

In 1997 the media reported throughout the world that a sheep named Dolly had been cloned from the genetic material of another sheep in the Scottish Highlands by Ian Wilmut and Keith Campbell. Wilmut and Campbell's experiment provided evidence for the hypothesis that every cell in a living creature not only holds all of the genetic information about the body but can actually reproduce the entire living system. The cloning of Dolly was one manifestation of the advances science has made in recent decades in understanding the genetic code and in providing answers to questions about the mystery of life itself.

Genetic research has so far brought us revolutionary new methods for treating illness, ways of identifying a criminal from a single hair left at the crime scene, and important information about where we came from. The human genome contains 100,000 genes: the collection of all the information about the traits of an individual. These genes are packed into twenty-three pairs of chromosomes, which reside in the nucleus of every cell in the body. In total, there are about 3 billion individual bits of information in the human genes. Scientists say that we humans share among us as much as 99.9 percent of the 3 billion bits of information coded in our genes. Thus, the individual differences between human beings account for only 0.1 percent of all this information: human beings are a lot more alike than they are different! It is this tiny variation in the totality of our genetic code that accounts for the diversity among individuals.

Yale University researchers Kenneth and Judith Kidd have compared genetic material from human populations of different regions of the world in a search to uncover our genetic diversity. Such changes occur due to mutations and are then preserved by the process of natural selection, where mutations that improve the individual's ability to thrive in its environment are favored in the individual's progeny. For example, the residents of an isolated village in Italy have been found to have heart disease rates that are surprisingly lower than those of other regions in Italy, despite their rich diet. On Réunion Island in the Indian Ocean, people have a natural immunity to multiple sclerosis. On the other hand, on Tristan da Cunha, an island in the south Atlantic, a full one-third of the population suffers from asthma. Then there are the long-known genetically favored diseases: sickle cell anemia affecting blacks, and Tay-Sachs disease occurring among Jews of Ashkenazi descent. Genetic studies of various populations around the world led the Kidds and other scientists to conclude that *Homo sapiens* originated in Africa 200,000 years ago, and that humans began to spread around the world only as late as 100,000 years ago. In 1988 scientists announced to a surprised world that genetic research has proved that all people on Earth are the descendants of a single woman, whom they appropriately named Eve.

We've managed to take genetic material from cells of living things and cause it to replicate in many ways to form genetically altered strains. We now have strawberries that have been genetically improved to increase shelf life; we have cows that give more milk because of laboratory-improved genes; we raise pigs whose livers are genetically altered to make them similar to human livers, for use in transplants. The Human Genome Project, begun in 1989, is constructing a chart of the entire genetic makeup of the human race. The $3 billion worldwide effort has already revealed many genes that make individuals susceptible to various cancers and other illnesses as well as a wealth of information about people's genetic properties; and by the project's completion, expected in 2005, we will know every gene and be able to match it with all of a human being's characteristics controlled by the gene. But despite the tremendous advances we have made in understanding genetics, no one has yet been able to construct the large, complex genetic molecule, DNA. It seems that the chemistry of life is infinitely more complicated than the "usual" chemistry of electrons, atoms, and small molecules.

Having surveyed the discovery of the structure of DNA and its properties—along with the role DNA plays in the life of cells and individuals and entire species—and having seen how DNA stores and manipulates tremendous amounts of information (3 billion separate bits for a human being) and uses the information to control life, we are left with one big question: What created DNA? Could this be a "molecule" in the usual, chemical sense of the word, or something far beyond chemistry? Could this microscopic computer of life be the result of atoms randomly reacting with each other chemically? Could DNA have formed in the same way as carbon dioxide, where two oxygen atoms meet and bond with a carbon atom and a specific amount of energy is released? Or are we witnessing here something so wondrous, so fantastically complex, that it could not be chemistry or random interactions of elements, but something far beyond our understanding? Was it perhaps the power, thinking, and will of a supreme being that created this self-replicating basis of all life, a molecule like none other—one that no one can replicate in no matter how complex a laboratory? How do we define something that can stretch, split along its middle, and clone itself—as a chemical molecule or as a living thing?

WE ARE NOW WITNESSING a revolution in the biological sciences. Not a single day passes without a discovery about genetics and DNA and the role of genes and chromosomes in our life. New genetic drugs are made for combating and preventing disease, and genetically engineered species are being brought to life. We are able to manipulate DNA in the laboratory—to rewrite the genetic code to our own specifications. We are even able to use all we've learned and clone living beings, something that was the fabric of science fiction only a few decades ago. But we have not been able to *create* life in a bottle. Our most advanced laboratories have not produced DNA from raw chemicals.

In the years that followed his groundbreaking codiscovery of the structure of DNA and his award of the Nobel Prize, James Watson did not rest on his laurels. He continued to work in genetics and became president of the prestigious Cold Spring Harbor Laboratory, where much research in genetics and biology in the United States takes place. James Watson is the founding director of the Human Genome Project.

To honor his great achievements, President Clinton awarded James D. Watson the National Medal of Science at the close of 1997. Watson's codiscoverer of the structure of DNA, Francis Crick, went off in another direction. If DNA is so complex that it cannot arise by chance, and if it is indeed the only molecule that can produce living things that ingest food, move about, and reproduce themselves, then how did it materialize on Earth in the first place? We will next encounter the theory Francis Crick took up to explain this apparent impossibility.

5 Panspermia, Byurakan, and the Meteorite from Mars

IN 1907 THE SWEDISH chemist Svante Arrhenius developed a theory about the origin of life. Life didn't originate on Earth, Arrhenius said. Simple living organisms, consisting of one cell, traveled from world to world throughout the universe in the frozen state of a spore. These tiny conveyors of life were propelled by radiation that pushed them away from a star and into empty space. The frozen organisms traveled interstellar space for eons, until they reached another solar system and settled on one of its planets. There, they came alive and reproduced, and over millions of years developed into higher life-forms. This theory became known as the panspermia hypothesis.

Arrhenius based his panspermia theory on what he thought had occurred here on Earth. Terrestrial microorganisms, he believed, were sometimes wafted into the stratosphere by high winds. Once these organisms were blown to very high altitudes, some of them would be ejected completely out of the planet by electrical forces. And once they were gone, they would travel the solar system and beyond. If this could happen on Earth, it probably happens everywhere in the universe, and therefore interstellar space may be full of small frozen organisms traveling around and fertilizing distant planets with life originating elsewhere in the universe. The panspermia hypothesis may be used to attempt to explain the origin of life on Earth by presupposing its existence elsewhere in the universe, where presumably it had more time to develop. This supposes that DNA can travel from one location in space to another by some viable way that protects it on its long journey.

In the 1960s Carl Sagan was interested in the panspermia hypothesis and its implications about the existence of life elsewhere in the universe. Sagan knew that balloon studies had confirmed that microorganisms can be found at great heights above Earth, even well into the stratosphere, confirming part of Arrhenius's idea. Sagan then tried to build a mathematical model that would determine whether it was possible for such organisms to be ejected beyond Earth's atmosphere into space, and what would be their likely fate. He described his results in his 1966 book, *Intelligent Life in the Universe*.[10] Sagan's model assumes that an organism that somehow escapes Earth will be subjected to two forces and that its fate would depend on the ratio of these forces. The first force, p, is the radiation pressure from the Sun, which will push the organism away from the Sun and into interstellar space. The second force, g, is the gravitational force of the Sun, which pulls the organism toward the Sun. When $p = g$, the organism stays in interplanetary space. If g is the greater of the two, the organism is doomed to fall into the Sun. But when p is greater than g, the pressure of the Sun's radiation is stronger than the Sun's gravitational attraction. In this case, the organism will sail on into open space, propelled by the force that is the difference between the radiation pressure and gravity.[11]

Based on these considerations, Sagan computed the size of organisms that can escape Earth. These are organisms with a radius between 0.2 and 0.6 microns (a micron is one-thousandth of a millimeter). Some bacterial and fungal spores as well as viruses fall within this size range and could conceivably escape the solar system, if they are able to leave Earth's atmosphere. Since the net force of the Sun's radiation minus its gravity continues to push on the escaping organism, it will eventually reach very high speeds. Within months it will reach the orbit of Jupiter, within years that of Neptune, and within a few tens of thousands of years such a spore could reach the nearest star. Within a few hundred million years, the bacterial or fungal spore that left Earth could transit the Milky Way. Sagan's coauthor, I. S. Shklovskii, points out that such an itinerant organism will have about the size and mass of a grain of the cosmic dust that permeates our galaxy. It may join such a cloud and become a part of it. At any rate, the movements of small, light particles such as these are governed by random interactions with other small particles, and the result is Brownian motion. This is the phenomenon of random walk, where a particle meets other particles suspended in the same medium, and the random collisions that result propel the particle in a random and completely unpredictable way. The motion is like that of a smoke particle in the air, sometimes drifting in one direction, then suddenly shifting to another random direction. The erratic random walk will make an organism's trip last longer than if it moved in a straight line.

A spore with a radius smaller than two microns will, because of the relative magnitudes of p and g for the Sun, *enter* our solar system if it came close to it, rather than leave it. When such an organism encounters Earth or another planet on its way to the Sun, it can enter the planet's atmosphere and eventually work its way down to the surface. If such an organism came to us from another solar system after a trip of tens or hundreds of thousands of years and then reemerged from its interstellar sleep, it could, in principle, replicate itself and set foothold in Earth's environment. Making some assumptions about the density of interstellar clouds that cause collisions with potential spores, as well as the time it has taken for life to develop on Earth, Sagan computed that life on Earth could have been seeded by an organism that had traveled here from a planetary system as far as six thousand light-years away from Earth.

Sagan and his coresearchers then turned their attention to the question of whether a transstellar traveler of this sort could actually survive its trip. Will the DNA of a tiny organism remain intact over thousands of years of interstellar wanderings? It was known even in Arrhenius's time that some spores could withstand very low temperatures for long periods of time and still maintain their ability to spring back to life once the conditions were right. Laboratory experiments have also shown that some organisms can survive long periods of time in vacuum conditions similar to those that prevail in space. It is possible, however, that organisms would not hold up well against the boiling away of atoms that can occur in a vacuum when they are exposed to such an environment for many thousands of years.

The biggest challenge to the panspermia hypothesis, however, comes from the problem of radiation in space. Within the environment just outside Earth's protective atmosphere, ultraviolet and other radiation from the Sun is relatively strong. It has been computed that in the neighborhood of Earth, a microorganism would not survive for more than a day after leaving the atmosphere and die from the Sun's ultraviolet rays. Assuming that the microorganism is somehow protected from the ultraviolet radiation, high-energy X rays and protons coming from the Sun would kill the spore within a few weeks. This problem does not exist in areas far away from the Sun or from other stars, or in the vicinity of stars dimmer than the Sun—those emitting radiation at lower intensities. In open space, however, a spore would encounter cosmic radiation. Assuming that the spore is as resistant to such radiation as the most radiation-immune organisms on Earth, it has been calculated that in open space, away from the radiation of stars, such an organism could withstand the cosmic radiation for 100 million years.

These assumptions and calculations, therefore, allow for the possibility of spores to have originated on planets that are as far from a star similar to the Sun as Jupiter or Saturn are from the Sun, or a planet in close orbit around a dimmer star. Since spores would probably be destroyed before reaching Earth as they encounter radiation from the Sun, the panspermia hypothesis does not look promising as a theory for the seeding of life on Earth by interstellar spores. However, the possibility of seeding life on a satellite of one of the Sun's distant planets does exist. Thus life might have sprung up on Jupiter's moon Europa or Neptune's moon Triton by some interstellar traveler. But even here, yet another problem appears.

Since the distances involved are so tremendous and the sizes of planets are so infinitesimally small within the cosmic scale, the *chance* of any single spore from another solar system arriving on a planet of another system is virtually zero. For the panspermia hypothesis to work, billions of tons of spores must be ejected in our galaxy over billions of years to produce one successful seeding. The panspermia theory rests on shaky ground.

With these conclusions in mind, Carl Sagan left for the Byurakan meeting in September 1971. An international conference on Communication with Extraterrestrial Intelligence (CETI) was to take place at the Byurakan Astrophysical Observatory in Soviet Armenia. To this distant location traveled the Americans Frank Drake and Philip Morrison, and scientists from Russia, Britain, Canada, Hungary, and Czechoslovakia. On the American side, the conference was organized by Carl Sagan, who was one of the key discussants at the meeting. Britain was represented by Francis Crick.

Crick came to Byurakan because he was interested in ideas about life in outer space that were going to be discussed at the conference. Crick started to believe in the possibility of life in outer space as a result of his work on the discovery of DNA. He believed that the initial formation of the DNA molecule was a unique event in the history of the universe. At the conference, he used the example of a deck of cards leading to one particular and rare configuration of cards. He further believed that Earth was too young for such a rare event to have occurred here.

According to scientific discoveries, the oldest fossils of organisms on Earth are about 3.5 billion years old. This would mean that life started here only about a billion to a billion and a half years after the formation of the planet. Crick thought that this time span was too short for the development of the incredibly intricate DNA structure. The panspermia hypothesis offered a way out. If DNA is so complex that it could not have originated on Earth, then perhaps it came here from some other planet, an older one where it would have had the necessary time to develop. Crick formulated his hypothesis about the extraterrestrial origin of DNA in joint discussions with Leslie Orgel of the Salk Institute in California while they were both attending the Byurakan meeting. Carl Sagan presented opposing views. Based on his computations of the radiation doses absorbed by organisms traveling interstellar space, Sagan argued that the panspermia hypothesis was very unlikely to hold. Instead, Sagan rejected Crick's analogy between DNA's appearance on Earth and the rare sequence of cards. In Carl Sagan's view, DNA—or something like it with a *different* chemical composition but fulfilling the same function—could have been independently produced on extrasolar planets.

On Earth itself, we see examples of panspermia. Any visitor to Hawaii is impressed by the lush vegetation and abundant, exotic wildlife. The islands are a veritable paradise with tropical plants and trees of so many kinds and bird varieties that are not seen elsewhere in one place. But millions of years ago, soon after the volcanoes broke through the surface of the Pacific Ocean and became islands, there was no life on them. Soon, the trade winds brought with them seeds of plants and trees and ferns and other vegetation from other tropical islands lying thousands of miles away, as well as from Asia and the Americas, lying farther away still. Eventually, birds and other animals arrived. On Earth, at least, living things can travel large distances to inhabit a new place, and over millions of years the place can develop its own rich ecosystem from species originating elsewhere. Is there a way that this can happen in space as well?

ON DECEMBER 27, 1984, Roberta Score, a scientist employed by the National Science Foundation, discovered a small rock in a place called the Allan Hills ice field in Antarctica. Score was in Antarctica searching for meteorites. Since meteorites are hard to find on the ground because they look, on the surface, like ordinary rocks and blend into their surroundings, the NSF concluded that the best place to look for meteorites was the barren ice fields of Antarctica, where natural Earth rocks are rare. When she first lifted up the potato-sized meteorite, Roberta Score thought it was a usual stone meteorite. But as she looked at it more carefully, she noticed something strange: the rock was green. Her meteorite was taken to the United States with the others found by the team, and nothing happened to it for a decade as it sat on top of a table somewhere in the back room of a lab.

Score's meteorite came to the attention of scientists who were aware of data sent to Earth from NASA's *Viking* missions to Mars in 1976. These data provided the exact composition of Mars's atmosphere through which the *Viking* craft were flying. When the scientists obtained Score's meteorite in 1993, they came to the exciting conclusion that the meteorite, now named Allan Hills 84001 (abbreviated ALH84001) had to have originated on the Red Planet.

We now believe that 15 million years ago an asteroid hit Mars with such force that rocks on the surface of the planet were blasted into space. One of these rocks was the 4.2-pound Allan Hills 84001. The intense heat that was generated as the object blazed outward through the Martian atmosphere sealed bubbles of the Martian air inside it. When the gas in ALH84001 was analyzed, its composition agreed exactly with the data sent to Earth from the *Viking* spacecraft. From the evident effects of cosmic radiation on this rock, scientists estimated that it had wandered around the solar system for 15 million years, fell to Earth, and sank into the Allan Hills ice cap about thirteen thousand years ago. Ice floes below the surface slowly moved the object upward until it reached the surface, where it was found by Score in 1984.

When Mars became a planet, it had water on its surface and an atmosphere with pressure close to that on Earth. We know that water flowed on the planet's surface in rivers and in and out of lakes and oceans. We still see the riverbeds that were carved on the surface of the planet over millions of years. After a billion years, Mars dried up. It lost its water and its atmosphere. Today the atmospheric pressure on the planet is seven millibars, which is the pressure on Earth at an altitude of twenty miles above the surface. Ablation has caused the atmosphere of the planet to disappear—being smaller than Earth, Mars did not have the strong gravity necessary to hold on to its precious gases and liquid water for more than a billion years after its creation. The loss of the atmosphere made temperatures on the planet swing widely: more extremely so than on Earth's most arid deserts. Today, the temperatures on Mars can differ by as much as thirty degrees over distances of a few feet above ground. Mars does not have the water, air, or stable temperatures needed to sustain any form of life. But 3.6 billion years ago, there was still water under the surface of Mars, and forms of life could have survived there. This is when ALH84001 was blown into space and began the 15-million-year journey that would eventually land it on Antarctica.

In 1994 Allan Hills 84001 was handed over to David McKay of NASA's Johnson Space Center. McKay assembled a nine-member team including scientists at NASA, Stanford University, the University of Georgia, and McGill University in Montreal. The team endeavored to determine whether there were any signs of life on the meteorite. The rock was believed to have formed under the Martian surface 4.5 billion years ago, when the solar system was created out of the planetary disk that surrounded the newborn Sun. The meteorite was close to the Martian surface 3.6 billion years ago, when underground water might have still sustained life on Mars. The scientists wanted to see whether signs of such microscopic life might be evident on the small rock in their hands.

McKay's research group subjected the meteorite to a battery of sophisticated tests. Allan Hills 84001 was bombarded with laser beams to separate different chemical compounds it might contain. The team discovered polycyclic aromatic hydrocarbons: organic molecules. These were similar to those found in coal or gas. When microorganisms die, such organic compounds are created. Could the compounds in the meteorite have been the remains of dead Martian organisms? We know that carbonaceous chondrites are meteorites with organic matter from outer space. However, ALH84001 was not a carbonaceous chondrite; it was more like a stone: a rocky meteorite—but it had these organic molecules inside it. A powerful electron microscope revealed tiny mineral deposits inside the meteorite. A mineral called magnetite was found, a compound of iron and oxygen. This compound is known to be excreted by terrestrial organisms. Could the magnetite in the meteorite have been produced by Martian organisms? A mineral of sulfur, called pyrrhotite, was found as well, and is known to be produced by bacteria and microbes. On Earth, such compounds are not known to occur without some life-related processes. Finally, microscopic structures were found in the meteorite: tiny tubes and channels that could have been carved out by living organisms. McKay's team noted the great similarity that the microstructures bore to ones made by tiny bacteria found on Earth.

Critics were quick to point out that none of the phenomena inferred from the analysis of ALH84001, each one by itself, constituted positive proof of Martian life. Other meteorites contained organic matter, the minerals in ALH84001 could have been created by natural forces that had nothing to do with life, and the microscopic tunnels found inside the rock could have formed as a result of chemical reactions and water activity rather than life.

McKay's team agreed but pointed out that the evidence, when taken as a whole, could not be dismissed. The researchers felt the implication of the study about the possibility of life on Mars was very strong. The news was announced to the world in 1996 that scientists might have found evidence for early life on Mars.

What may have added to the belief that life was present on Mars at one time while ALH84001 was still on that planet, 3.5 billion years ago, was evidence from *Viking* and other sources that water was present and flowing on the Martian surface at that time. If water was there, scientists believed, life could have been present as well. Moreover, the earliest evidence for life on Earth dates from about the same period. We know from fossilized microorganisms that life was present on our own planet as early as 3.5 billion years ago, and there are chemical telltales of the activity of microscopic life on Earth as early as 3.9 billion years ago. Ironically, it was this finding, based on scientific dating techniques, that led Francis Crick to his belief in the panspermia hypothesis. Crick had subtracted 3.9 billion years from 4.5 billion years (the estimated age of our solar system) and got a difference of only 600 million years. According to the renowned geneticist, DNA could not have formed on Earth within a span of merely 600 million years—much more time would have been required for the random collisions of atoms of carbon, hydrogen, phosphorus, and so on before an immensely complex molecule such as DNA could be formed. Thus, he reasoned, life must have developed elsewhere and somehow implanted on Earth. The announcement in 1996 about traces of possible life on a rock from Mars quickly revived the panspermia theory. Articles began to appear in scientific journals, with titles such as: "Are We Martians?"

Allan Hills 84001 was not the only piece of Martian rock ever found on Earth. There are twelve such meteorites that scientists can identify with high confidence as having arrived here from the Red Planet. However, ALH84001 is of the right age, according to some researchers, to provide evidence of life. Mars was wet during some of the time the rock lay close to the surface of the planet. The whole concept of what happened to ALH84001 on Mars and on its long trip to Earth until its discovery in 1984 seemed to answer any contentions against the panspermia hypothesis in an elegant way. Carl Sagan would have been greatly impressed by the arguments, since back in 1971 he had reluctantly played devil's advocate at Byurakan against a theory he would have loved to endorse, given his well-known desire to prove someday that life must exist elsewhere in the universe.

Sagan's main argument against panspermia was that prolonged exposure to intense radiation in open space would destroy any form of life within hours, days, or weeks. But Sagan and his colleagues at Byurakan had considered organisms that were small enough to soar to the highest levels of Earth's or another planet's stratosphere and then gain enough energy to depart into space. Small, unprotected organisms flying by themselves into space would indeed suffer from radiation hitting their sensitive DNA, eventually destroying it completely. But ALH84001 was a completely different story. Here, whatever organisms became embedded in the rock while it was on Mars remained protected from radiation once they were deep inside the rock's natural protective radiation shield. Could such an extraterrestrial rock have seeded life on Earth?

Everyone has seen shooting stars. A shooting star, or a meteor, is what will become a meteorite if it survives the entry into Earth's atmosphere and makes it all the way down to the ground. A meteor is visible because it shines brightly and looks like a star streaking through the night sky. The reason we see a meteor is that the object gets heated to very high temperatures due to friction with the atmosphere. The heat produced is so great, in fact, that most small objects entering Earth's atmosphere burn on their way to the ground. And here is where another problem arises with the panspermia theory. When ALH84001 was blasted into space, it heated up so much that it trapped bubbles of Mars's atmospheric gases on its molten surface—allowing us to identify its origin. But the same temperature that sealed in the signature gases of the Red Planet might have also killed any living organisms so that what scientists have now found in the rock are only traces of former life, if that's what they are. The same heating process certainly took place when ALH84001 entered Earth's atmosphere. So while we can assume that a meteorite might, under the right circumstances, protect an organism and its DNA from the radiation in space, and while the organism may be assumed to be frozen throughout its journey and might thus survive the trip in a state of suspended animation, we are still faced with the problem of the sterilizing heat produced during departure and arrival.

Can this problem be solved? Possibly, the life-laden rock thrown into space from a planet somewhere could be composed of matter that resists heat: for example, asbestos. Another possibility hinges on the ratio of surface area to volume. A potato-sized rock such as ALH84001 can heat up to high temperatures, but what about larger objects? The Moon itself was once a piece of Earth until a gigantic collision resulted in its departure into orbit around us. It's easy to imagine objects larger than ALH84001 but smaller than the Moon flying out of a planet due to some collision with an asteroid and carrying deep inside them—where they could be protected from both radiation and heat—living organisms. After traveling through interstellar space for thousands of years, such objects could land on a planet around a different star, where life could be seeded from the frozen spores hidden deep within the stellar courier.

EARTH'S ATMOSPHERE IS CONSTANTLY bombarded by cosmic rays from outer space. This radiation breaks up atoms of nitrogen in the upper levels of our atmosphere, forming atoms of the radioactive element carbon 14. The radioactive carbon works its way down to ground level and permeates everything in our environment. So while most of the carbon in the world is of the nonradioactive kinds (carbon 12 and the less common carbon 13), a small but fairly constant fraction of all carbon in circulation is radioactive. Now, living things constantly absorb carbon from their surroundings. They ingest it as food, breathe it and absorb it from the air, and so the proportion of radioactive carbon in living things is the same as it is in the environment. Not so once an organism dies. At the moment a living thing is deceased, no more carbon enters it from the surroundings or leaves it. The radioactive carbon already in the remains of the organism continues its natural decay into nonradioactive elements, and in the long run, there will be no radioactivity left. But after a number of years, only some fraction of the radioactivity will have diminished. After 5,730 years, half of the radioactivity present at the time the organism was living will have vanished. This is so since 5,730 years is the half-life of carbon 14. Similarly, after 11,460 years, only a quarter of the original radioactivity will remain, and so on. This predictable characteristic of radioactive carbon gave rise to the radiocarbon dating technique, invented in the 1940s by W. F. Libby.

The method was refined in 1993 by Dr. Minze Stuiver of the University of Washington and his colleagues, who undertook to calibrate radiocarbon dates against the most precise method of dating trees: counting rings on a tree trunk. By performing radiocarbon analysis of pieces of wood with known ages and known dates of cutting, a sophisticated comparison was made of the dates of the trees with the radiocarbon dates. By the early 1990s, the errors in measurement of radiocarbon dates had been reduced to plus or minus 10 to 20 years. Radiocarbon dating is now an accurate scientific technique that has been used in a variety of applications. These have ranged from proving that the Shroud of Turin was produced from flax harvested in the 1200s rather than the time of Jesus, to dating Joshua's conquest of Jericho to 1580 B.C. and dating Neanderthal bones to 100,000 B.C. For older dates, especially in determining geological ages of rocks, similar radioactivity analyses (potassium-argon and other techniques) are used.

In early 1998 Dr. A. J. T. Jull of the geosciences department at the University of Arizona announced the results of an analysis that made new use of the radiocarbon technique. Jull had obtained small pieces of ALH84001 and burned them in special laboratory reaction vessels. He thus obtained the results of the combustion of the hydrocarbons found in the meteorite and isolated the carbon that was part of these organic chemicals. Jull also separated other samples of carbon from ALH84001—these ones obtained from the inorganic calcium carbonate mineral that is part of the Martian rock itself. Then Jull looked at each carbon product and analyzed the ratio of the radioactive isotope, carbon 14, to the other two isotopes, carbon 12 and carbon 13. Jull's results were amazing. The organic carbon from the meteorite had a ratio of carbon 14 to carbon 12 and 13 that was identical to the ratio found in terrestrial carbon, but different from the ratio found in the carbonate, presumably from Mars. Jull's conclusion was that any organisms within ALH84001 had to have gotten into the meteorite while it lay on the Antarctic ice field. Other studies reported at the same time also brought into question the origin of any telltale of Martian life found inside the meteorite.

The consensus among scientists in early 1998 was that nothing conclusive could be said about the mystery of ALH84001. Scientists agreed that the meteorite most probably came from Mars but could not agree on much else. Each one of the findings that had originally led researchers to the belief that traces of ancient life were evident on the little Martian rock had come to be shadowed by doubt. Everyone agreed that the answer to the big question of whether life existed on Mars at some point in its past would have to be answered by future analysis of Martian rocks, either brought to Earth or analyzed by a space vehicle on the surface of the planet. Planned missions to Mars in the first decade of the next century would, we hope, bring us the answer. When this happens, and if the answer is positive, we may finally know whether DNA is unique—at least within our own solar system—or whether other forms of the genetic code molecule are possible.

Do Cosmic Couriers Exist?

Could interstellar panspermia work, moving life from one solar system to another? Most extraterrestrial bodies—meteorites, asteroids, comets—originate within our own solar system. Asteroids live in the asteroid belt between the orbits of Mars and Jupiter, where another planet would have formed according to Bode's law. This law (also called the Titius-Bode law) describes the distances between planets in our solar system and the Sun. Take the series: 0, 3, 6, 12, 24, 48, 96, 192, . . . and add 4 (since Mercury is about 0.4 AU from the Sun) to all its numbers, giving: 4, 7, 10, 16, 28, 52, 100, 196, . . . When these numbers are divided by 10, they give: 0.4, 0.7, 1, 1.6, 2.8, 5.2, 10, 19.6, . . . These are roughly the distances, in astronomical units (AU), of the planets from the Sun. For Mercury, the actual distance is 0.39 AU; for Venus, 0.72 AU; for Earth, 1.00 AU; for Mars, 1.52 AU. Then there is a gap right where Bode's law predicts the next planet, at 2.8 AU from the Sun. After that, we have Jupiter at 5.2 AU, Saturn at 9.5 AU, and Uranus at 19.2 AU (for Neptune and Pluto, the law does not offer a good approximation).

Where the gap in the series was found, G. Piazzi discovered the asteroid Ceres in 1801, and later many other asteroids were found at the distance of 2.8 AU from the Sun, in what is now known as the asteroid belt. The asteroids are debris that remained from the creation of the solar system. As the protoplanetary disk surrounding the newborn Sun coagulated at roughly the distances given by Bode's law, at this distance of 2.8 AU there was just too little matter to produce a new planet and what remained were the would-be building blocks for a planet. The asteroids orbit the Sun within this belt, and some of them roam our solar system, having originated in the belt. Ceres, the largest asteroid, has a diameter of 1,025 kilometers. Pallas, Juno, Vesta, and two hundred other known asteroids have diameters of several hundred kilometers, and there are thousands of smaller asteroids with diameters of a kilometer or less.

Beyond the planets, outside Pluto's orbit, lies another belt. This one is called the Kuiper Belt and it consists of icy bodies left over from our solar system's birth. At times, one of these bodies may loosen itself from its orbit and fall into a path leading toward the Sun—to become a comet. The Kuiper Belt supplies short-period comets: those that orbit the Sun with a period of less than two hundred years. The asteroids and comets that roam our solar system can bring matter from one location to another. But can they transport life?

Farther out in space, thousands of AU from the Sun, lies another collection of bodies circling our Sun. This is the distant Oort Cloud, named after the Dutch astronomer Jan Oort, who in 1932 also discovered the existence of dark matter in the galaxy. The Oort Cloud is believed to have formed early in the life of our solar system, as clumps of matter were scattered away from the center of the solar system by the gravitational force exerted by the large planets Jupiter, Saturn, Uranus, and Neptune. Objects in the Oort Cloud produce comets with periods of thousands of years. These comets are icy objects that are part of the cloud; sometimes a gravitational perturbation from the outer planets sends such an object on its way to orbit the Sun—and the period of a complete revolution is in the thousands of years.

As stars themselves revolve around the massive center of our galaxy, at times they get closer to one another. When this happens, Oort Clouds—assuming other stars have them, too—can intermingle. Then, a rocky, icy object born in one solar system can travel to another. In addition to the Oort Clouds, scientists now believe that rogue comets roam interstellar space, unattached to any solar system. These are comets that were pulled away to a distant planet in a solar system, and the "slingshot effect" catapulted them out into space, where inertia keeps them going ever outward and into the vast open regions between stars. According to a recent estimate, the space between us and the nearest star system, Alpha Centauri, could contain as many as 50 *trillion* such rogue comets. It's further believed that there may be as many as 6 trillion trillion (6 with 24 zeros) interstellar comets in the Milky Way galaxy. With such a large number of travelers from one solar system to another, it is possible that the chemical building blocks of life may have traveled through the galaxy, and the panspermia hypothesis may be viable after all, assuming DNA can survive the hardships of such long trips under extreme conditions.

6 Asteroids, Volcanoes, and Nemesis

LIFE ON EARTH—and possibly on any planet in the universe—is fraught with danger. It is this danger of extinction, or danger of life never even having a chance to evolve, that made Frank Drake and his SETI colleagues consider a factor in Drake's equation that would account for the precariousness of life on any planet: the *L* factor. This factor measures the length of time any civilization might exist. The *L* factor should account for a number of different elements that endanger the evolution and longevity of life, and of intelligent life in particular, on any planet. Here we account for the self-destructive tendencies of our species, as evidenced by the production and maintenance of weapons of mass destruction, and the possibility that such inclinations might be present in other advanced civilizations. But the dangers to life anywhere in the universe come from many directions. The universe is a violent place, as every astronomer knows. There are explosions and bursts of intense radiation occurring in space all the time, and on any planet there are dangers of collisions with objects. The key question here is: How likely is it that the conditions somewhere in space will be benevolent enough for a long enough period to allow life to evolve and be sustained over time?

In July 1994 comet Shoemaker-Levy 9 crashed into Jupiter in what became the most widely observed event in astronomical history. For several days every professional telescope in the world was trained on the giant planet. As the comet broke into pieces when it encountered the massive resistance of the Jovian atmosphere, its debris smashed into the surface of the planet with such force that impact scars the size of the entire Earth could be seen on the face of the planet even through small telescopes. Right after the comet broke up, huge fireballs of hot gas and dust rose high above the Jovian landscape and Jupiter's atmosphere darkened. A large impact spot remained permanently on the surface. The questions on everyone's mind were: Is this a rare event or a common cosmic phenomenon? And what really happens to the environment of a planet that gets hit by a visitor from outer space? Can life escape such natural disasters and continue, or must all life, everywhere in the universe, eventually come to an end by a fiery explosion? And, will Earth ever face such a fate?

Until twenty years ago, few scientists suspected that Earth was in similar danger of being hit by objects from space. This is surprising, given that a glance at the Moon's battered landscape should convince anyone that space is a dangerous place. The Moon gets hit more frequently than Earth because it has a very thin atmosphere and objects don't burn as they enter it, as do most objects entering

Earth's atmosphere. Smaller objects that fall into Earth's environment burn up completely before reaching the surface. Also on Earth, geological activity, which is lacking on the Moon, causes erosion of the surface as tectonic plates move against each other and cover up many old impact craters, hiding them from sight. But Earth did get hit many times in its past by large objects from space. And these impacts caused great catastrophes.

In 1978 Louis and Walter Alvarez of the University of California at Berkeley were in the Italian Apennines studying geological layers dated to the boundary between the Cretaceous and Tertiary periods (abbreviated as the K/T boundary), 65 million years ago. The two scientists discovered something they had not expected: a thin layer of rock that was rich in the element iridium. This element is a relative of platinum and is extremely nonreactive; it is not even soluble in aqua regia. Iridium rarely occurs naturally on Earth; but iridium has been found in meteorites. The fact that the iridium was concentrated in such a thin, well-defined layer prompted the Alvarezes to conclude that there was a strong possibility that the iridium came from outer space and arrived here via an asteroid. The scientists then computed from the density of the iridium layer they had discovered—and using the assumption that the layer was distributed evenly throughout Earth's surface—that the total amount of iridium in such a global layer had to have come from an asteroid with a diameter of ten kilometers. What struck the scientists most was that the time period of the geological iridium layer and the asteroid's putative impact coincided with the time put forward for the extinction of the dinosaurs.

But if there really was such an asteroid that hit Earth during the K/T boundary, then where were the remains of the asteroid? Where was the crater created by the collision impact of a body of such great mass? Such a crater would have to be huge. But by the time of the Alvarezes' discovery, no such crater had ever been discovered on Earth. And therefore, for all everyone knew, the iridium layer could just as well have been a coincidence. The whole new theory about the cause of the extinction of the dinosaurs by the impact of an asteroid on Earth lacked a smoking gun.

Then in 1991 a gargantuan crater was discovered. It lay right under Mexico's Yucatán peninsula. A layer of the signature iridium linked this find directly with other layers that had been discovered around the world during the intervening years; and all of these, together with the original find in the Apennines, were dated to the K/T boundary 65 million years ago. It became clear that Earth, just like Jupiter in 1994, was once the target of a giant invader from space. This unwelcome visitor was dubbed the Great Extinctor. From the size of the crater, scientists determined that the magnitude of the asteroid's impact was equivalent to that of 5 billion Hiroshima-sized atom bombs, or 100 million megatons.

Within seconds of the impact, a great fireball rose high above the ground and into Earth's atmosphere. Temperatures shot to searing levels. But some animals and plants, especially ones that could hide and were not close to the point of impact, did survive the heat. It was the aftermath of the great explosion that caused most of the damage to our planet. Clouds of soot and dust rose high up and spread all over Earth's atmosphere, resulting in a dark cover

over the entire planet. The darkness lasted many months and prevented the Sun's radiation from reaching the surface of the earth. The immediate high temperatures from the explosion quickly gave way to freezing cold that lasted for years. Since the heavy black cover over Earth prevented the normal warming of the planet by the Sun's rays, the annual cycle of rising and falling temperatures was gone and instead our planet remained very cold. Lack of light prevented plants from carrying out photosynthesis and most of them died. As everything on the ground froze over and the plants died, so did the dinosaurs and most other living things on Earth. What life did survive the calamity, and how such life continued to evolve into the living planet we have today, is still a mystery.

The scientific analysis of the K/T explosion and its effects on life on Earth led the Alvarezes to another unexpected theory. During much of the Cold War, the superpowers were both guided by a policy that assumed that a nuclear war was winnable. From calculations of the force and the damage that would be unleashed by a single nuclear explosion—information obtained through nuclear tests over a period of forty years, above ground at first and later underground—Russian and American nuclear scientists came to believe that atom bombs destroy by shock wave and fire, as well as by radiation. But what scientists now began to understand was that there was yet another effect of large explosions. If many nuclear bombs were exploded all at the same time, as in a nuclear war, the damage would not be contained. Burning cities would produce so much soot and dust that when all of these products of the initial explosions rose into the atmosphere, Earth's temperatures would drop to dangerous levels and the Sun's radiation would not reach the ground. Scientists named this effect, which was of the same nature as the K/T disaster, nuclear winter. The nuclear winter hypothesis played an important role in convincing the superpowers that nuclear war was probably not survivable and would bring us the fate of the dinosaurs. Thus the nuclear winter theory, which was promoted by scientists on both sides and brought to the attention of world leaders, helped bring about the end of the Cold War.

It seems that it may be difficult to sustain life on Earth—or on any planet—over very long periods of time. The very process of the formation of solar systems from protoplanetary disks to planets creates the elements that threaten the continuity of life. Asteroids, comets, and other hunks of matter in space seem to be integral parts of a solar system in the same way that planets are. So while the panspermia hypothesis that bodies coming in from outer space can spawn life is iffy at best, the destructive power of these same bodies, and the threat they pose to the development and continuity of life, is very real. There are various levels to the danger.

ON JUNE 30, 1908, a loud explosion was heard in the villages of the Tunguska region in central Siberia. People who looked up saw a huge fireball in the sky as everything turned red and they felt the temperature rise dramatically as in a giant fire. The wide-ranging forest fires that ensued devastated an area of two thousand square kilometers. But because of the region's remoteness, scientific expeditions to learn the cause of the disaster did not take place until seventeen years later. Today we know that the culprit was a rocky asteroid the size of an office building. As the intruder from space decelerated when it met the resistance of Earth's atmosphere, it heated up and ultimately exploded eight kilometers above the surface, spreading fire over the entire region. We know that the power released by the asteroid's explosion was the equivalent of fifteen megatons of TNT—a thousand times more energy than was released by the bomb dropped on Hiroshima.

Scientists estimate that an object the size of the Tunguska asteroid, roughly fifty meters in diameter, hits Earth once every three hundred years. If such an object should fall on a city, it would completely destroy it. But since the impact points on Earth are random, such asteroids have so far exploded over sparsely populated regions, such as Tunguska. Assuming that 10 percent of the planet's surface is densely populated, scientists have calculated that a Tunguska-like object would devastate a populated area once every ten thousand years. But there are other dangers from asteroids or comets of this relatively small size. A Tunguska-sized asteroid hitting the ocean would create tidal waves that would devastate large populated areas on the coasts of the ocean. A recent simulation by supercomputer has shown that even a relatively small asteroid crashing into the Atlantic Ocean would produce a tidal wave so powerful that it would reduce the cities of the East Coast to wet rubble.

A larger asteroid, with a radius of one hundred meters, would produce the impact force of a hundred-megaton bomb and would devastate an entire continent. An asteroid with a diameter of one kilometer (0.6 miles) would produce the destructive power of a 100,000-megaton bomb and would wipe out life on an entire hemisphere. The Great Extinctor of the K/T era, ten kilometers in diameter, produced the destructive force of impact of 100 million megatons and destroyed life on a global scale. Scientists estimate that such disasters, of the largest magnitude, can occur roughly once in 30 million years.

There is a theory, not yet supported by any scientific evidence, that the Sun is not a single star but rather part of a double-star system, like most stars in our galaxy. According to this theory, the Sun's companion, named Nemesis, revolves around the Sun once every 30 million years. When its elliptical orbit gets Nemesis close to our Sun, Nemesis' gravitational field attracts comets and asteroids and sends them speeding toward the Sun and its planets. The theory asserts that this is how the Great Extinctor came to hit Earth and wipe out the dinosaurs and most life on our planet 65 million years ago.

Asteroid 1997 XF11

Though most asteroids stay in the asteroid belt, there is one group—called the Apollo asteroids—whose orbits cross Earth's orbit. There are over thirty such rocks in space whose paths cross that of our planet. One of these asteroids, Icarus, came within 4 million miles of Earth in 1968. In astronomical terms, this is considered a near miss. More recently, asteroids have been known to cross our path in space at a distance of less than 300,000 miles. Then on December 6, 1997, Dr. James V. Scotti of the University of Arizona discovered a new asteroid. This one, later named Asteroid 1997 XF11, was observed to have an orbit that could get it very close to Earth, and in March 1998 scientists announced that the newly discovered asteroid was on a course that would eventually lead it to within only 30,000 miles of Earth—closer than any body ever observed. The asteroid takes twenty-one months to revolve around the Sun, and it has been calculated that on Thursday, October 26, 2028, at 1:30 P.M., Asteroid 1997 XF11 will pass closest to Earth. Coming to within 30,000 miles of our planet—one-eighth the distance from Earth to the Moon—could pose danger of a collision. Later calculations by scientists at the Jet Propulsion Lab in California led to more optimistic estimates of XF11's closest passage point to Earth, varying from 54,000 miles to 600,000 miles. If it hit Earth, the asteroid—being a mile in diameter—could cause destruction on a global scale. The asteroid will be observed when its orbit brings it again to our neighborhood in the years 2000 and 2002, and better estimates of the danger of collision will be made then. In a twist of irony, one possibility for avoiding collision with the asteroid could be hitting it with a missile bearing a nuclear device, which, when exploded, could throw the asteroid off course. It is easy to see that with near misses from asteroids or comets, over a period of many millions of years, a direct hit is likely to occur.

Nuclear winter—not to mention radiation, explosion damage, and heat—can destroy any advanced civilization at war on a planetwide scale. Asteroids and comets from space—assuming other solar systems have such remnants of their early days roaming their own interplanetary space and perhaps the space between star systems—pose yet another danger to an extraterrestrial civilization. On Earth, we know that at least one planetwide disaster did occur because of an asteroid. Ironically, intelligent life may never have arisen on Earth in the first place had the dinosaurs not become extinct following the K/T explosion and its aftermath. And there is yet a third danger that lurks in any planet that can support life. We know that to support life, planets must be geologically active. A geologically dead body such as the Moon lacks many of the elements necessary for life to evolve. But the very same geological activity that helps a planet come alive can also destroy it. This dangerous geological activity is vulcanism. Volcanoes are a natural phenomenon on active planets. Jupiter's satellite Io is so active geologically that huge plumes of volcano smoke can be seen in pictures from the Hubble Space Telescope. Frequent and intense volcanic activity can destroy life in a way that is very similar to nuclear winter.

WHEN THE DINOSAURS died 65 million years ago, there were no humans on Earth and we therefore have no record of the level of destruction caused by the Great Extinctor—except for the geological and paleontological findings. How might a similar catastrophe affect *people*? This question was on my mind in 1991, when I was doing research on statistical methods to improve the radiocarbon dating technique. I used an advanced mathematical method called the bootstrap to obtain a narrow confidence interval for radiocarbon dates of prehistoric events. As part of this project, I wanted to visit the site of the most devastating natural calamity in human history—the place where archaeologists had found the remains of an ancient civilization that had mysteriously disappeared from the face of the earth around 1630 B.C. Among the archaeological finds were jars containing grains of barley left by the people who died in this great disaster. These grains were used in radiocarbon dating.[12]

I climbed to the top of the rim of the huge volcano on the island of Santorini. By a terrace with a good view of the caldera below, I read a sign: TRESPASSING FOR CUSTOMERS ONLY. Walking south for half an hour, I rounded a sheltered area behind a small hill and was again facing the enormous blue caldera below me, this time looking north. It is from here, the southern rim of the volcano, that one can grasp the sheer size of the crater. It was a clear day, and yet I could hardly see the northernmost town of Ia, directly across the caldera. The amount of earth and rocks blown into the air that day or over several days in the seventeenth century B.C. must have been enormous. All around me were large black basaltic rocks. They looked as if they had fallen out of the sky, landing scattered over the yellow-gray earth. The same northerly wind must have been blowing then, when the eruption occurred (as was in fact proved by studies of ash in deep-sea cores), and the rocks and ash and pumice caused a hellish devastation as the wind swept them in the air to the south and east. A thick layer of ash covered the fields of Crete, seventy miles to the south. Modern science has shown that coverage by volcanic ash to a depth discovered on Crete would make such arable land unsuitable for human use for many years. The strong winds carried the enormous cloud of ash southeast toward Egypt, where it caused darkness that wouldn't lift for days. As I was thinking of the cataclysm that took place so long ago, the sun was beating down and it was very hot and very dry, but I finally made it to the entrance of the village of Akrotiri.

A short time after the archaeologist Spyridon Marinatos first started excavating south of Akrotiri in 1967, his team recovered incredible finds. An entire ancient city lay under the ash and pumice, just like Pompeii. But this one was seventeen centuries older. Two-storied houses with running water and flushing toilets lined streets and a city square. Inside these houses Marinatos's team found elaborate frescoes and beautiful vases crafted by artisans of a civilization that later disappeared suddenly and without explanation. Marinatos's work received worldwide media attention, and the *National Geographic*, which in 1967 published beautiful pictures of the frescoes and art treasures recovered at Akrotiri, described his work as the unearthing of Atlantis.

In 1932 Marinatos was excavating at the site of Amnisos, the harbor that served the great palace of Knossos on Crete. Marinatos was working at a relatively high point near what had been the harbor, excavating the remains of a Minoan villa. What he discovered struck him as bizarre. The villa seemed to have been pulled right off its foundations by some mysterious force. As Marinatos began noticing similar signs of unexplained destruction at other sites on Crete, he became convinced that the Minoan civilization did not come to its end through war or other conventional ways. The Minoans were the victims of a sinister natural force of unprecedented scale.

In 1939 Marinatos published an article in the English periodical *Antiquity*. In his article "The Volcanic Destruction of Minoan Crete," Marinatos proposed his theory that a violent eruption of the Santorini volcano was the cause of the destruction of the Minoan civilization on the island of

Crete. According to Marinatos, the eruption was accompanied by violent earthquakes that caused great tidal waves. The Minoan settlements on Crete were destroyed by the tidal waves and by the large amounts of volcanic ash deposited on the land, making it unsuitable for cultivation.

But Atlantis is not the only legend we have about the total and sudden destruction of a human civilization by natural forces common in the universe. This one was confirmed with reasonable certainty thanks to the work of Marinatos and others. There are other legends that we may attribute to similar natural disasters. The flood of Deucalion, the flood of the ancient Sumerian epic of Gilgamesh, and Noah's flood may all be folk retellings of tidal waves that occurred as a result of earthquakes or volcanic eruptions—or from the impact of asteroids crashing into the sea. The human experience on Earth is just a moment in the cosmic timescale, so the fact that human civilizations have not yet completely disappeared does not prove that life can survive indefinitely. Within the hierarchy of cosmic accidents, the larger ones occur more rarely and the smaller ones more frequently. However, if we lived, as a species, as long as the dinosaurs had lived on Earth—a period of 180 million years—we too could be destroyed by a planetwide disaster. The *L* term in Drake's equation is elusive. We simply don't know how long a civilization can last, but we do know it is not forever—which has implications about the precariousness of life-sustaining conditions on any extrasolar planets. For even if we find such planets within the habitable zone around their stars, where water may exist in liquid form and where the temperatures may be ambient, frequent volcanic activity or impacts of asteroids or comets may prevent life from being sustained there.

The anthropic principle states that the existence of life makes it necessary that conditions favoring life as we know it exist. By this principle, if the electron had a mass just slightly lower or higher than it has, life couldn't exist. If the carbon atom had a different size, molecules that give rise to life couldn't exist. And if our solar system were any different from what it is, we wouldn't be here. This principle asserts that everything in the universe must be exactly the way it is, or else we wouldn't exist. In a sense, the anthropic principle is related to the idea that Earth is the center of the universe, a view held until the 1500s, when Copernicus argued that it was false and that the Sun was the center of the solar system. Then, a century later, Galileo's observations of the satellites orbiting Jupiter were among the first pieces of evidence that Earth was not central to everything. There are, however, properties of Jupiter and of our Moon that scientists have used to build arguments that seem anthropic in their nature.

Jupiter is a giant planet, 318 times as massive as Earth. Because of its great mass and the fact that its orbit lies farther away from the Sun than ours, Jupiter is believed to be a great protector of Earth. When comets or asteroids stray from their own orbits and pose a danger of crashing into Earth and destroying life, these intruders are more likely to move toward the massive Jupiter than toward Earth. Thus Jupiter is a great attractor, whose gravitational pull protects us from anything that may pose a danger to life on Earth. To a smaller degree, Saturn, with a mass equivalent to ninety-five Earths, also acts as our protector from flying bodies in space. According to this theory, comet Shoemaker-Levy 9 did not crash into Jupiter by chance—it had to go there because of Jupiter's size, which attracts comets and asteroids. The theory about the role of Jupiter and the other large planets in protecting Earth from space debris suggests that if these planets did not exist, Earth would be hammered by comets and asteroids at a much greater frequency than we now believe it is. One estimate is that instead of once in 30 to 60 million years, a large crash of an asteroid or comet would occur once every 100,000 years if we didn't have the large planets' protection. Thus life could not have developed if these giant planets did not exist.

When considering the chances of life elsewhere in the universe, some scientists claim that we need to consider the probability for the existence of such large planets to protect a life-bearing planet like Earth. Such an argument may be seen as a reflection of the anthropic principle: if we have a Jupiter and a Saturn, living things on another solar system must have them, too.

A similar theory, advanced in 1993 by the French astronomer Jacques Laskar, claims that life would not have developed on Earth if we did not have a moon. Sometime in the early days of our solar system, a large object hit the young Earth. The impact was so great that a piece of Earth was launched into space. Its giant mass created a gravitational pull so strong that the body eventually compacted into a spherical shape, forming the Moon. Because the Moon, being so massive for a piece of space rock, remained close to Earth, it started orbiting Earth, each body feeling the gravitational pull of the other. It is the force of the Moon's gravity that creates the tides on Earth, and some scientists have contended that without the tides, life would have stayed in the oceans rather than climb out onto the continents washed by the tides. But Laskar's theory went even further in assuming that the Moon was necessary for life to develop on Earth.

The Moon is larger than the planet Pluto, and not much smaller than Mercury. Yet the Moon is very close to Earth—a couple of days' flight by spacecraft, rather than the months or years it takes to fly to other planets. As a consequence, the Earth-Moon system can almost be viewed astronomically as a two-planet subsystem within our larger solar system. Being so close to Earth and having a sizable mass, the Moon is gravitationally "locked" into its orbit around the earth, so that it cannot rotate. This is why we always see the same face of the Moon. According to Laskar, it is this locked-together gravitational dance the Moon performs around the earth that provides Earth with its stability of axial tilt. Earth revolves around the Sun not perpendicularly but at a tilt. The axis of rotation is at an angle of 23.4 degrees to the perpendicular. It is this tilt of Earth's axis of rotation around the Sun that gives us the seasons. When Earth's northern hemisphere is facing the Sun during daytime, it is summer, and when the southern hemisphere faces it, it is winter. When the Sun's rays fall perpendicularly on the equator, it is the spring or fall equinox.

But Earth's axial tilt is not constant: it varies over a period of thousands of years from a minimum of 22.0 degrees to a maximum of 24.6 degrees. This phenomenon is called precession, and it is responsible for a component of the long-term movement of stars around us. The North Star,

Polaris, was in ancient times a different star. Three millennia ago, the star above our North Pole was Beta Ursae Minoris. Because of the rotation of the Earth's axis, that star of antiquity slowly slipped away from its place right above Earth's North Pole, so that now the star right above our North Pole is another member of the Ursa Minor constellation: Alpha Ursae Minoris, the star we call Polaris.

According to Laskar, the Moon provides Earth its relative stability of axial rotation. If we didn't have the Moon orbiting us so closely, locked with one face toward us, Earth's axis of rotation around the Sun would swing wildly—much more than it currently does. Laskar maintains that without the Moon, Earth's axial tilt would swing wildly from an angle of zero degrees to an angle of eighty-five degrees to the perpendicular. This would cause wild variations in the seasons: sometimes there would be no seasonal variation at all (when the tilt angle was zero), and other times winter and summer would be so extremely different in temperature that winter would cause widespread freezing and summer would scorch the earth. It is hard to say how true this scenario might be, or whether this theory falls under the anthropic argument: since the Moon is here, we must have it or else life could not be sustained.

The problems raised by our apparent need for both Jupiter and other large planets and for a moon seem to imply that the probability for the existence of the right conditions for life in outer space is low. For what is the chance of us finding another solar system that has the right kind of star, the right kind of planet—a small, rocky world with all the necessary elements for life, orbiting the star at the right distance so that water can exist in liquid form—

and also having a moon just like ours and outer large planets protecting the planet from space debris?

But in 1997 D. Williams and J. Kastings of Pennsylvania State University conducted an extensive computer simulation of conditions for life on planets that do not possess a moon like ours. Their results, published in the September 1997 issue of the journal *Icarus*, showed that many such planets without a moon could still maintain conditions favorable to life. Possibly, what we have on Earth in terms of the relative stability of the axial tilt is not an absolute necessity. And maybe Laskar's research has been interpreted too literally, and falls within the realm of the anthropic principle. It's nice to have a moon, but life could be sustained on a planet without one.

As for the large planets with strong gravitational pull necessary to deflect rogue comets and asteroids, the discoveries of Michel Mayor and Didier Queloz and those of the other planet hunters around the world have already given us proof that gas giant planets such as Jupiter do exist elsewhere in the universe. In fact, astronomers have been finding only planets the size of Jupiter. And if Jupiters abound in the universe, then the small, rocky planets like Earth—once they are detected—should be amply protected from space debris.

While dangers to life both on Earth and in any extrasolar planetary system are significant, it is hard to assess their real impact on the development and continuity of life. Dangers arise both from outside—the possible impact of asteroids or comets on any planet that may harbor life—and from within: the planet's natural geological activity, causing volcanic eruptions, some of which can be violent enough to create a nuclear winter effect. In addition, swings in the climate due to an instability of the tilt of the axis of a planet with respect to its plane of revolution around its star can also be a hurdle to life. To what extent these concerns are anthropic in nature, and to what extent they are real, is an open question. New discoveries in space, however, have already gone part of the way to allay our concerns.

AMONG THE DISASTERS attributed to the Santorini eruption, some have counted the Ten Plagues of Egypt and the parting of the sea. The assumption has been that the eruption caused the three days of darkness described in the Bible, and that the remaining plagues were secondary effects of the same event. The parting of the sea has been attributed to the tidal wave that resulted from the Santorini eruption. Marinatos himself cared only about Atlantis and was never a proponent of the ten plagues and parting of the sea hypotheses. Marinatos fell to his death on site at Akrotiri in 1974, and the cause of death was reported as a stroke leading to his losing his balance on the high ledge on which he was standing, making him fall.

IT WAS A SATURDAY NIGHT, and Panayotis and his wife and daughter dressed well. They wore the festive costumes of their island: embroidered white shirts, elegant blue trousers or skirts. "Tonight, we go together," Panayotis had said to me after breakfast that morning. I tried to dress as well as I could—traveling with a small bag didn't allow for many possibilities. At seven I met the three of them in the hallway. We got into the red minivan, the three of them up front and I in the back. Panayotis took the dusty, bumpy road from the town of Naxos to the mountain village of Filoti. We drove past cultivated fields and barren hills where a few sheep grazed. The three of them were happy, singing Greek songs.

In half an hour, we entered the well-lit square of the village. As we turned, a bearded man in his thirties, standing on the sidewalk by his car, ran into the street, smiling and waving. Panayotis stopped next to him and the two started talking excitedly in Greek. We parked and joined the man and his wife and daughter, everyone kissing each other and shaking hands with me. "*Synadelfos sas*," said Panayotis as he introduced me to his friend. "Your colleague." Andreas was a professor of archaeology, and so in that sense we were colleagues, since Panayotis and his family knew I was a professor.

Andreas didn't speak any English, but his twelve-year-old daughter, Eleni, was a good translator. We entered the restaurant, everyone in a festive, relaxed, and happy mood. First came the *mezedes*, the hors d'oeuvre that tourists to Greece go crazy over: cucumber-yogurt salad, fish-eggs salad, chopped octopus. Then came pizza and bottles of retsina and red wine. And the conversation flowed. The two families were apparently very close friends, and I felt privileged to be invited to such an intimate outing.

"I was in Santorini," I said, when Andreas finally turned to me with a question. "I was amazed to see Marinatos's grave—I hadn't known that he was buried right there on site, in Akrotiri, where he fell." Andreas smiled. "Oh, yes, Marinatos. He was a wonderful archaeologist," he said. "But he didn't just fall. Marinatos was murdered." I looked at him stupefied. *Murdered?* Eleni was translating word for word as he was speaking. "You didn't know?" he asked. "The people didn't like all these theories that were coming out because of his work. You know, about the parting of the sea and about the plagues of Egypt. The workers Marinatos had with him on site were very religious people. They were also primitive and superstitious. To them it was all the work of God, not a volcano. They had to kill him, to stop the blasphemy."

The Evolution of Intelligence 7

IN A VIOLENT UNIVERSE, fraught with dangers from outside a planet, as well as from within, how does life evolve once the DNA molecule has been created? What is the progression from DNA and the simplest life-forms such as viruses and single-celled creatures to fish, amphibians, reptiles, and then on to mammals, the most developed of which are primates, including the most advanced: human beings? How does the evolution come about, and can we reasonably expect it to occur elsewhere in the universe?

Six hundred million years ago, at the beginning of the Cambrian period when the Precambrian era came to a close, a giant volcanic eruption occurred in China. Studies of this eruption in 1998 led to a revolution in our understanding of the development of life on Earth. Two teams of paleontologists have found distinct traces of ancient marine animals preserved in phosphate. These animals were more complex than the sponges and jellyfish that had been known to have developed around this time and were, in fact, the first known fossils of two-sided animals—that is, animals with two symmetric sides, like the human face with its two eyes, two ears, and mouth and nose that can be split symmetrically down the middle. The new findings established that early life-forms on Earth were abundant and diverse, in contrast with the conventional chronology, which maintained that the "Cambrian explosion," the big bang of biology that created the great diversity of life-forms on Earth, had occurred around 40 million years later than the date of the new finds.

Using radioactive dating techniques, the revolutionary new findings were dated to about 600 million years ago, and scientists now conclude that multicellular organisms on Earth may have evolved as early as half a billion years before the Cambrian explosion. Radiocarbon dating techniques do not work in establishing dates so far in the past. Here, a related method was used, which relies on the decay of uranium to lead. Knowing the half-life of uranium and measuring the proportion of lead in ancient deposits of uranium found in the rocks in the area covered by the lava from the eruption, scientists were able to arrive at the date of 600 million years for the two-sided organisms whose remains they had uncovered.

In *The Lives of a Cell* the late renowned biologist Lewis Thomas asserted that the uniformity of all life on Earth, as evidenced on the microbiological level, leads science inevitably to the conclusion that all life on Earth descended from a single living cell.[13] From this one cell we got our looks and characteristics as creatures mutated and evolved over billions of years, molded by the forces of natural selection. All living things from grasses and trees to dolphins and wolves and humans share genes as well as enzymes and cell structure. How did this evolution take place, eventually transforming single-celled organisms into human beings who can reason and create?

Single-celled organisms are already beyond the mere DNA strings of viruses. These living creatures can move around in the water. The single cell has cilia, or hairs, that allow it to move by wiggling the hairs, or it may move around by squirming. These very basic living things respond to their environment. A single-celled organism—a cell with a nucleus with genetic matter and with mitochondria, enclosed in a shell of protein—responds to being poked with a needle by trying to escape the intrusion. It also responds to chemical changes in the environment, moving around to seek a more favorable environment when the present one deteriorates. Single-celled organisms even respond to light. It is incredible to believe that such simple forms of life already possess the properties of much more complex beings: the ability to seek food, the ability to escape danger, and the ability to strive to reproduce themselves.

As we go up the ladder of life from simple single-celled organisms to multicellular ones, we encounter more diversity in living systems. The Taylor Valley in Antarctica is one of the most desolate places on Earth. When the explorer R. F. Scott discovered it in 1903, he called it "the valley of the dead," but new explorations of the area have revealed simple life-forms. These were found under the ice, in rocks and in the soil. Here, researchers were surprised to find millions of microscopic plants, single-celled animals, and nematode worms. Scientists considered the system they had found the simplest ecosystem on the planet. The nematode worms were the only multicellular animals found here. The way this ecological system works is that the bacteria in the ground ingest algae from a lake, then the nematode worms eat the bacteria. Researchers consider this area an excellent model for closed ecological systems, and they have been studying the food chain by tracing how carbon travels from the water and air to the algae, then from the algae to the bacteria, and finally from the bacteria to the worms. This system has revealed much about how the first multicellular organisms developed in the soil and were able to feed on single-celled organisms. Three and a half billion years ago, this ecological system was the only one on Earth—carbon ingested by algae, leading up to bacteria and nematode worms.

But at some point in the development of life on our planet, higher forms of life came about. As more advanced animals develop, their cellular structures become specialized. Thus a fish or an amphibian will have a system of cells devoted to movement, a sensory system to receive information from the environment, a digestive system, and so on. For the worm, as an example of a simple life-form, there are circular muscles and longitudinal muscles that allow the worm to shorten itself and lengthen itself and twist around. The muscle system even allows the worm to swim.

The diversity of life on Earth is immense. But to truly grasp this amazing variety of creatures, we must also consider animals and plants that have become extinct. Among all the extensive families of living creatures, we find development of the various systems of the body. When we look at families of living and extinct creatures, we find an interesting phenomenon: all these families, in their own way, have developed forms of intelligence. The essence of intelligence is to learn and anticipate from the environment. An intelligent animal can learn from its environment, then it can predict outcomes and, finally, change its position within the environment to adapt to it in the best possible way. The remarkable aspect of intelligence on Earth has been the constancy of intelligence from species to species. A prey learns how to escape its predator, be it fish, bird, or mammal. And a predator learns how to pursue its prey in the most effective way, whether the predator is a shark or a tiger or an extinct carnivorous dinosaur. In this sense, even the single-celled organism that can wiggle around and try to escape being poked with a pin exhibits a form of intelligence. This is not to say that it can think or create, but it does possess a primitive way of responding to its environment.

As organisms developed from single-celled creatures that existed for 3 billion years into multicelled organisms with specialized cells, nerve cells (also called neurons) appeared. This process of evolution culminated with intelligence. Intelligence has evolved as part of the sensory system of living organisms. Nerve cells are branched like a tree. In the center of the cell we find the nucleus with the genetic material, as in other cells. From the cell body extend a number of branches called dendrites. These dendrites are in close proximity to those of other cells, and that is how connections (called synapses) are made for the transmission of information. Nerve cells may have a long branch that transmits information, called an axon. The nerve cell integrates all the information received from neighboring nerve cells, and then a signal is sent through the axon. Ions of potassium and sodium inside and outside the axon produce electrical currents, which transmit the information. In a typical process, the work of the nerve cells and their communications with each other through the dendrites will produce a decision. Then the decision is translated into a command sent through the axon to a muscle to result in a particular action. This type of process is exhibited by primitive forms of life as well as by the most advanced.

The first development of nerve cells in living creatures appeared in the form of sensory organs: eyes, ears, nose. The nerve cells in a central location in the head of the organism receive messages from the sensory organs and process the information. Then the nerve cells give commands to the muscles to respond to the sensory stimuli and move. Through the evolution of the nerve cells, the brain and intelligence came about.

The brain, as the central organ of the sensory system in an organism, began as a collection of nerve cells with their dendrites and axons. These controlled the animal's movements, senses, and most bodily functions. As more advanced organisms evolved in all branches of the animal kingdom, the brain became more sophisticated and developed increasing levels of intelligence. A mouse exhibits traits of intelligence as it learns to find its way through a maze. A fox learns from its parents how to hunt its prey. A fish learns how to find food, and a beaver learns how to build a dam; the latter activity is one we consider a higher intelligence since it involves not simply adapting to the environment but actual construction.

Charles Darwin wrote in *The Descent of Man:* "No one, I presume, doubts that the large proportion which the size of man's brain bears to his body, compared to the same proportion in the gorilla or orang, is closely connected with mental powers." Thus the nineteenth-century developer of the theory of evolution and natural selection already saw a possible connection between the relative size of the brain and the intelligence of a living thing. Today, we call the index of the relative size of the brain of a living creature the index of cephalization. This index is defined as the total brain size divided by the two-thirds power (meaning squaring and taking the third-root) of body size. This is an adjusted measure, which scientists have found to work much better in explaining brain power than a simple ratio of brain size to total body size.

Paleontologists have been able to estimate the size of the brain and the total size of the body of animals that are now extinct based on the size and shape of bones that have been recovered. Thus, in addition to the data on currently living animals, we also have a good idea about the cephalization of the dinosaurs. Here, a surprising fact emerged. Dinosaurs are not "small brained," as the popular belief goes. Scientists have found that the statistical relation between body and brain sizes is constant for both the extinct members of the reptilian class and for living reptiles. Ten dinosaurs that were studied exhibited the same cephalization as living lizards. Fish, amphibians, reptiles, and mammals all have a characteristic cephalization level. Of these, the mammals' level is the highest. But differences do exist within each of these classes of animals.

The degree of cephalization of primates is the highest in the mammal class, where it is highest of all animals. For all animal classes, we see an increase in cephalization through time. Relative brain size was small for all living things in the period 40 to 60 million years ago. In the next period, 25 to 55 million years ago, relative brain sizes for all animals roughly doubled. Then in the period 5 to 20 million years ago, there was yet another doubling of brain size, and another one from then on to the present era. We see, therefore, that brain sizes, and with them intelligence, have been systematically increasing through evolution. Within each class of animals, scientists have posited an "expected" relative brain size, which is a reflection of the cephalization level for the entire class. For each animal, a measurement is taken of the deviation from the expected level. Here we find that within the primate class, smart monkeys have cephalization levels that are higher than expected. The largest deviations are for humans, dolphins, monkeys and apes, and Neanderthals and other forms of ancient humans.

The people involved with the search for extraterrestrial intelligence (SETI) are interested in the existence of intelligent life in outer space. They would not be happy if, someday, we should find an extrasolar planet inhabited by cows. And some biology experts have asserted that intelligence as we usually think of it—the ability to think and plan and solve a crossword puzzle and argue convincingly at a board meeting—is an evolutionary fluke. These scientists use the argument that of all the classes of the animal kingdom, only one, that of humans, has intelligence. Hence, the argument goes, intelligence has a small probability of occurring elsewhere.

But everything we know about evolution and natural selection from studies of animals and their brains indicates that this is not the case. It is true that a monkey can't type *Hamlet* except by chance and that a dog can't play chess. On the other hand, we see from the fossil records and from animal studies that there is a clear and consistent progression over millions of years toward larger brain sizes as compared with body sizes and an increase in intelligence as life evolves to higher levels. Recent findings about changes in the environment confirm the theory that life leads to intelligence. Lizards on an island with a closed environment have been exposed to new dangers in the last few decades as predators have been introduced to the environment. To researchers' surprise, within a few generations the lizards have learned to climb trees to escape the new predators. This quick adaptation to the environment over such a short period of time gives a strong indication of how intelligence can evolve quickly—much faster than scientists had expected. In *The Number Sense*, cognitive scientist S. Dehaene describes experiments on rats.[14] The rodents were taught to press two levers different numbers of times (for example, four times on lever A and twice on B) in order to receive food. These experiments continued with increasing levels of complexity and involved counting light signals as well as tones. They have shown that the rats had a sense of numbers and counting. The author claimed that people's ability to understand mathematics is an evolution of the simpler sense of numbers that is present even in animals such as rats. Some forms of intelligence, including basic mathematical ability, are evidently present in other living things.

While human beings are the only animal we define as intelligent—able to reason, able to solve complex problems, and able to communicate at a high level—the remarkable findings in the realm of the evolution of intelligence in the animal world is that lower forms of intelligence have developed in parallel in all classes of animals. A bird can adapt to its environment and learn to improve its search for food and safety in the same way that a hare does or an insect. Such findings lend strong support to the belief that the evolution of intelligence is something inherent to life and—given time—would occur in any life-supporting environment.

Recently, a systems theory has gained favor with many scientists. In the early 1970s Stephen Smale—a professor of mathematics at the University of California at Berkeley and one of the world's leading experts on differential equations—began to study the relationship between predator and prey populations. By then, Smale had already achieved the greatest distinction in the world of mathematics: in 1963 he was awarded the Fields Medal—the mathematician's equivalent of the Nobel Prize. Stephen Smale earned this high honor for solving the Poincaré conjecture, one of the most famous unsolved problems in mathematics. Now Smale was looking for a mathematical equation that would capture the nature of the relationship between two populations. It was known that there was some kind of cyclical variation in the numbers of moose and wolves, as well as other predator-prey combinations, over time. But no one had been able to describe or understand the nature of this relation mathematically. Some years the moose were dominant, with few wolves around. Other years the situation reversed itself inexplicably, with lots of wolves but few moose. Then again, a reversal occurred seemingly out of the blue. Smale had a hunch that the high-level mathematics of which he was master might offer an answer to this mystery.

A differential equation is a mathematical formula that contains elements of change: typically it is a function of a variable (here, the present level of an animal population) and a rate of change in the variable, called the derivative of the variable (here, estimated from the rate of change from year to year of the level of a population). Smale proposed a particular differential equation and assessed its parameters from the wolf and moose population data. He also studied another case: that of the Canadian lynx. Since this animal, prized for its fur, had been hunted for over a century before conservation laws in Canada were imposed, there was much available data on this population and its prey. In both cases and in subsequent studies, Smale found that his equation surpassed expectations in describing the mathematical relationship between predator and prey. The numbers were not random: there was an exact mathematical formula that predicted the numbers of the two populations and how these numbers must change from year to year. The predator and the prey populations were performing a mysterious dance around each other: when one was up, the other was down and vice versa, in a perfect rhythm of life. In a sense, both populations needed one another. The predator fed on the prey, and the prey relied on the predator to keep its own population lean and healthy, since the predators tended to get the ailing and old members of the prey population. There was a mathematical harmony between the two populations and, indeed, among all populations of animals that inhabited a given area. The discovery that a mathematical formula could describe natural phenomena such as the relationships among several populations of animals was considered a significant contribution to our understanding of biology.

The predator-prey research made scientists see that in order to understand life, one must consider an entire system and not simply an individual. Life is not an issue of a single living thing. Living creatures all belong to a larger picture. One could not understand the life of an individual lynx, for example, any more than one could understand the life of a single cell in the body viewed as an independent living entity. Ultimately, we are all members of a larger group of life, which, in the limit, encompasses all of life on Earth. This hypothesis became known as the Gaia theory.

The Gaia hypothesis was proposed in the 1960s by James Lovelock, an atmospheric chemist at NASA. Lovelock analyzed the temperatures on Earth over a long period of time. He noticed that while the Sun's radiation has increased by 25 percent since life began on Earth, Earth's surface temperature has remained constant. Lovelock came to the conclusion that Earth should have been much warmer. There was something about the planet that made the temperature constant throughout such a long period of time. When the concentrations of oxygen and carbon dioxide and other gases were studied, it became evident that the processes of life were, at least in part, responsible for the planet's constancy of temperature. For one characteristic of a living thing is that it takes in energy and matter and produces waste products. A living planet, one that supports flora and fauna, would have an abundance of reactive gases. On Earth this gas was oxygen. A dead planet will, by the same token, contain an abundance of gases that were formed by chemical reactions that had been completed long in the past: the stable carbon dioxide, for example. The continual formation of oxygen by processes of life was keeping the levels of carbon dioxide low, therefore preventing the planet from overheating. Thus the planet supports life while at the same time this life provides the planet with the continuity of the same conditions that are necessary for life to exist.

Entire systems of life have their own collective intelligence. The best example of this is ants. A single ant has very few neurons and can hardly be expected to create much or to plan or analyze situations and make complex decisions. However, an ant colony can farm fungi, raise aphids as livestock, and use chemical sprays in battle against intruders. Ants exchange information with each other, and, in that sense, each one of them is like a single nerve cell that communicates with other nerve cells through the dendrites in the brain of an animal. Here, the entire system has its own collective intelligence, which it uses in the processes of the life of the entire colony.

Bees, with their sophisticated method of exchanging information, are another example of the existence of collective intelligence in a system of living things. A bee performs a complex dance, sending a message to other bees: "Clover can be found in a north-by-west direction, eight hundred meters away." When building a hive, the entire group swarms about, each bee performing an exact task, together building symmetrical polygons into a complete hive. Here, too, in terms of intelligence, each bee acts like a single nerve cell as in the brain of a higher-order animal.

Brain scientists have spent many years researching just what intelligence means. And a recent theory suggests that there are many different forms of intelligence. Some people think that a computer possesses intelligence, and, in fact, Garry Kasparov, the world's smartest chess player, was defeated in 1997 by an IBM computer. But does the computer really possess intelligence? A computer is a machine that is programmed to perform various kinds of mathematical operations: addition of binary numbers (that is, 0 or 1) and the comparison of two numbers being the two basic elements. Since the computer is so fast and since many processors can be linked in parallel, we get as a result the tremendous computing power of the machine. But what is it that separates the intelligence of an advanced animal from the raw computing power of a machine?

The computer doesn't care if you shut it off. A living thing will do anything in its power to avoid such a fate. The very nature of intelligence in people or animals is the ability to take actions that promote self-preservation, or preservation of the colony in the case of bees and ants. People and animals have emotions and feelings. In people, art and music are an intelligence, and while computers can be programmed to produce some forms of art or music, they lack the natural desire to do so. Computer scientists and experts on artificial intelligence claim that someday soon there will be computers that are no different from the human brain. These computers will be able to create and feel and want. As of now, however, I can quietly shut off my computer without fear of retaliation when I turn it on again. So computers, while they have taught us much about processes of thinking—for example the solution of complex mathematical problems or games of chess—do not hold the key to intelligence as we see it in nature.

The fact that intelligence has evolved separately in the various branches of the animal kingdom is proof that intelligence is not a fluke. It is a natural outcome of the evolution of living systems. Intelligence is brought about by the need of the individual and the society to preserve itself. Lower forms of life produce many offspring so that a few may survive. Animals whose survival is more assured require fewer offspring since their intelligence allows them to escape from danger. Ultimately, the evolution of a naked ape who comes into this world with no protection at all from the environment required that such a being be endowed with enough intelligence to build a shelter, harvest and hunt food with implements its society invents, and communicate with members of the society through a sophisticated language. The evolution of language is one of the most complex manifestations of intelligence people have exhibited—and one that sets us apart from other beings. Recently, researchers at UCLA have discovered the location in the brain responsible for a sense of humor and the ability to laugh. This area of the brain is adjacent to the one that controls language ability and communication, and the theory has been put forward that these two areas of the brain have evolved together as primates became more advanced, giving rise to early forms of humans and *Homo sapiens*.

Strong support for the hypothesis that intelligence has evolved separately in the various branches of living systems and that intelligence is an inevitable development in living beings' road to improved adaptation to their environment is offered by the theory of the evolution of birds. The origin of birds was, until recently, one of the greatest mysteries of biology. Fossil discoveries in the last twenty years, however, have helped scientists solve the mystery.

Birds are different from all other living creatures. They have feathers, toothless beaks, hollow bones, perching feet, wishbones, stumplike tailbones, and a host of other skeletal features that they share with no other living animals. Shortly after Charles Darwin published his theory of evolution in *On the Origin of Species* in 1859, a single bird's feather was discovered in limestone deposits in Bavaria. We now know it belonged to a birdlike creature that lived 150 million years ago: just before the Jurassic period gave way to the Cretaceous period.

Then in 1861 a skeleton was discovered in the same area in Bavaria of an animal that had birdlike wings and feathers, but with a long, bony tail and a toothed jaw—all uncharacteristic of birds we know today. This small birdlike creature became known as *Archaeopteryx*. Nine years later, at a meeting of the Geological Society of London, the first scientific discussion took place linking birds with dinosaurs. The assertion was that the 150-million-year-old fossil *Archaeopteryx*, the earliest known bird, was the link between dinosaurs and birds. Various treatises as well as criticisms of the theory followed over the next fifty years, arguing about the similarity of the anatomy of birds and the anatomy of dinosaurs inferred from the fossil record. The main criticism of the linkage theory was based on the fact that theropod dinosaurs—the closest family to birds—lacked clavicles, the two collarbones that in birds are fused together into the wishbone. Gerhard Heilmann, a Danish doctor and paleontologist, argued in 1916 that this lack of a clavicle in dinosaurs made it impossible that birds had descended from them, despite the remarkable resemblance between the fossil bones of the theropod dinosaurs and those of birds.

But in 1936 Charles Camp discovered the remains of a small theropod from the early Jurassic period. What struck the California paleontologist was that the skeletal remains he had uncovered included clavicles. Here was finally the missing link between dinosaurs and birds. In the last few years, many separate remains of theropod dinosaurs and early birds had been found, from Spain to China to Mongolia. These dinosaurs did include clavicles, as did the remains of early birds beyond *Archaeopteryx*. As a result, Heilmann's objections to the theory have been answered and the evidence is now considered overwhelmingly in favor of the theory about the origin of birds. Scientists have identified a large number of traits of birds and dinosaurs that are shared in common by both groups but not with any other group of living or extinct animals.

The *Velociraptor* was the first known dinosaur with birdlike characteristics. This animal had long, grasping arms and a swivel wrist. It chased its prey by running fast on its hind legs and seized the prey with strokes of the forelimbs. These forelimbs began to bear some resemblance to wings. The *Archaeopteryx* already had wings and feathers designed for flight. It had a shorter tail than the *Velociraptor*, since it began to move like a bird rather than like a lizard with a long tail. Flight began on trees. Some dinosaurs learned to climb up trees for safety from predators and to search for food. As these dinosaurs developed longer forelimbs and shorter tails, and as feathers grew on the winglike limbs, they could glide down from the trees, flapping their forelimbs in what were the first attempts to fly. Another hypothesis based on the fossil record maintains that some dinosaurs learned to fly as birds by running on the ground and flapping their feathered wings. The *Iberomesornis*, an ancient bird recently discovered in Spain, shows skeletal fusion in the direction of a small, aerodynamic body that is suited for sustained flight. Birds that developed later began to nest in trees, fly as the main method of moving around, and eventually lost the teeth that can be found in fossils of the earliest birds, which are close to their dinosaur ancestors.[15]

What is remarkable about this theory is that the link between birds and dinosaurs is actually a link between reptiles and birds. Reptiles are known to have small underdeveloped brains. As birds began to evolve from dinosaurs, their brains developed beyond the levels of those of reptiles, and so did their intelligence. Studies of cephalization levels have shown that bony fish have the lowest ratios of brain weight to body weight. The next group up is reptiles, which have higher levels of cephalization than bony fish. Interestingly, birds are much higher on this scale and their cephalization ratios are closer to those of mammals—in fact, species by species on a graph of cephalization levels, one finds birds intermingled with mammals and far above reptiles. On the top of the scale, one finds primates, including humans. Thus we see that the evolutionary process that brought reptiles through the extinct dinosaurs to birds has resulted in an increase in brain capacity and in intelligence. While the dodo bird became extinct because it lacked the ability to adapt to a hostile environment, many species of birds alive today are known for their remarkable intelligence. F. Nottebohm of Rockefeller University has carried out an extensive study of the neural basis for singing in songbirds. Nottebohm discovered in most species of birds he studied that the left side of the bird's brain is developed toward the singing ability. This specialization of part of the bird's brain is different, however, from development of the brain of mammals. Mammals have an entire part of the brain that does not exist either in reptiles or in birds: the neocortex. In humans and other primates, 70 percent of the neurons in the central nervous system reside in the neocortex. We, along with all other primates, have an entire area of the brain to separate us from birds. Still, one cannot but marvel at the fact that intelligence and relative brain size have clearly exhibited growth as reptiles evolved into birds and that a general forward direction in the evolution of intelligence seems to be a fact of nature.

While human beings are certainly unique and are the only species on Earth to have developed the full intelligence that allows us to communicate on all levels—to think, to analyze, to create—one cannot ignore the simple fact that other living creatures do possess lower forms of intelligence and that these intelligences seem to have evolved in a forward direction. Let's not forget, as well, that it has taken over a billion years for the simplest forms of life to develop on this planet, and over 4 billion years for them to develop into multicellular organisms, which appeared here only 600 million years ago. The first animal brains, very primitive collections of neurons controlling the movements and sensory ability of early creatures, appeared around that time. Then it took a full 400 million years for the first mammals to develop, around 200 million years ago. Our apelike ancestors did not come up on the stage of life on Earth until around 4 million years ago. And the age of *Homo sapiens* is measured in the thousands of years. When one considers how long this process of evolution has taken, one can see that the direction of increased intelligence is uniformly forward, and that intelligence has increased relatively quickly. Given enough millions of years from the time the DNA molecule arrives or evolves on a planet, intelligence will inevitably be the ultimate outcome.

8 Does God Play Dice?

AN IMPORTANT QUESTION comes up as we consider the development of life on Earth and the prospects for the emergence of life elsewhere in the universe: Are life systems, and other systems in the universe, random or determined? Randomness would imply that probability theory is the way to approach the question of the emergence of life in the universe. Determinism would have other implications about this issue. Did DNA come about by chance, or was it created in accordance with some mathematical prescription?

I came out of Professor Smale's course on differential equations with an unexpected appreciation for the order and regularity of life and amazement at how this order can be modeled by mathematical equations. It seemed that nothing was given to the whims of chance—the world, nature, the universe, all were programmed by precise mathematical statements of symbols and numbers that determined everything that happened and will happen again.

Kepler's equations governed the movements of the planets around the Sun; the laws of gravitation determined motion, speed, and time; and biological equations determined how species evolved, flourished, and adapted to their environment and to a world they shared with other species in a clocklike rhythm of rise and fall in strength and abundance. If a set of equations could describe the complex relationships among species, I wondered, then what could be said about the universe as a whole? Is there an intrinsic structure to nature—one that eludes the casual observer but suddenly emerges when one appeals to mathematics?

This question about the nature of the universe has its origins in the early decades of the twentieth century, when modern physics began to develop as a theory with far-reaching powers to explain the intricacies of the physical world around us. At that time, two important theories revolutionized this science. One was relativity—both the special theory and the general one. The other was the theory of quantum mechanics. Relativity was the work of Albert Einstein, and by 1917 he had finished his work on the general theory, having discovered special relativity a decade earlier. It is through Einstein's theories that today we understand the nature of space and time and gravity, with their implications about the limit of speed, black holes, the big bang, and the fate of the universe. Relativity is the theory of the large and the very fast (the speed of light being the ultimate constant in the universe), and in its very essence, the theory is *deterministic:* relativity gives answers to surprisingly complicated questions about the universe without any consideration of *randomness.* If a physicist knows the velocity, the mass, the force, and so on, then he or she can solve an equation and obtain a precise answer. Randomness, or chance, plays no part in this theory.

In the 1930s quantum theory was proposed. This was a theory of the very small: atoms, electrons, protons, and other entities that make up the microstructure of matter. Quantum mechanics was independently developed by the Austrian physicist Erwin Schrödinger (1887–1961) and the German physicist Werner Heisenberg (1901–1976). The first used equations, the other used matrices, but both obtained similar results describing the behavior of particles. One of the basic elements of quantum theory is the wave equation of a particle. The idea was that a particle the size of an electron is not only a particle—it is also a wave. The duality between particles and waves was proposed by French physicist Louis de Broglie. What is interesting about the wave equation one associates with a particle is that the square of the wave function is a *probability distribution*.

When one squares the rule describing the wave motion of an electron, for example, one gets the Gaussian probability distribution. Quantum mechanics explained to science the orbitals of the electrons in atoms. And these orbitals are all given as probability statements. An orbital is a region of space around the nucleus where the electron has an appreciable probability of being at any given instant. The famous Heisenberg uncertainty principle says that it is impossible for us to know with precision both the position and the momentum of an electron at any given time. Hence, all we can state about an electron in orbit around the nucleus of an atom is a probability distribution that tells us where the electron might be at any time. The orbitals are spacial probability distributions: they provide us with a region in space, a sphere, a dumbbell, or another kind of region in three-dimensional space where the electron might be.

Because of this inherent reliance on probability rather than on precise statements, quantum theory is essentially probabilistic in nature. This fact is in striking contrast with the purely deterministic nature of the theory of relativity, where one can determine the gravitational force, the speed, and even the contraction of time with great precision. And the connections between probability theory and quantum mechanics go much deeper than the orbitals of the electrons. Part of quantum theory is statistical mechanics. Here we describe the behavior of large numbers of small particles—say, all the molecules in the air in a room—and we get the Gaussian probability distribution: the normal, bell-shaped curve of errors, as it has been called, appears as the dominant force of the theory. For in order to describe the statistical behavior of a large number of particles, we must rely on statistical distributions. A statistical distribution—the distribution of a large number of items—serves naturally also as the probability distribution for a single item from the large collection. Thus the Gaussian, or normal, distribution is both a probability distribution and a statistical distribution. What is the Gaussian distribution?

Carl Friedrich Gauss (1777–1855), the great German mathematician who was believed to have derived many important results in mathematics long before anyone else, was also credited with the discovery of the normal law of errors. This law of probability theory states that in the limit, as many random factors are added together, the distribution of the sum of these random factors will look like a bell-shaped curve.

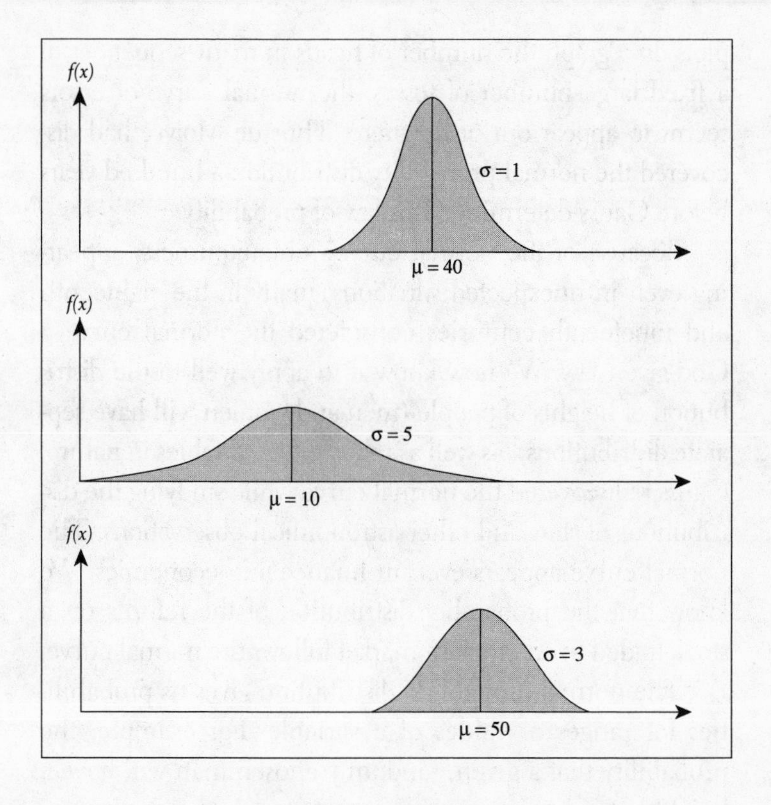

In 1909, however, the English statistician Karl Pearson discovered a pamphlet written sometime around the year 1720 by Abraham de Moivre, an impoverished mathematician, a Huguenot who emigrated from France to England to escape religious persecution and made his living as a coffeehouse consultant to students of mathematics. De Moivre had derived something we now call the central limit theorem, which demonstrates the convergence of the sums of many random factors to the mysterious bell-shaped curve. One example considered by de Moivre was the sum of heads appearing in a large number of tosses of a fair coin. If one plots on a graph the number of heads in many sequences of a fixed large number of tosses, the normal curve of errors seems to appear out of nowhere. Thus de Moivre had discovered the normal probability distribution a hundred years before Gauss determined this law of probability.

Because of the normal curve's ubiquitousness, appearing even in unexpected situations, many in the eighteenth and nineteenth centuries considered the normal curve a God-given law. We now know it to apply well to the distribution of heights of people (men and women will have separate distributions), as well as many other variables in nature. Gauss rediscovered the normal curve while studying the distributions of stars and other astronomical observations. The normal curve appears even in finance and economics. We know that the probability distribution of the returns on a stock traded in an efficient market follows the normal curve.

The normal probability distribution gives us probabilities for ranges of values of a variable. For example, the probability that a given, randomly chosen man will have a height between sixty-six and seventy-two inches is given as the area under the normal curve between these two values, as shown below. This probability is about 68 percent.

Intelligence scores are known to follow a normal distribution, although this fact has recently been abused in bad science with unsubstantiated statements about differences in intelligence levels among races. What the curve implies in general is that most values lie within a few standard deviations (standard units reflecting the natural variability of the specific data set) of the average, with smaller and smaller probabilities of falling further and further away from the mean in the "tails" of the distribution. The tails

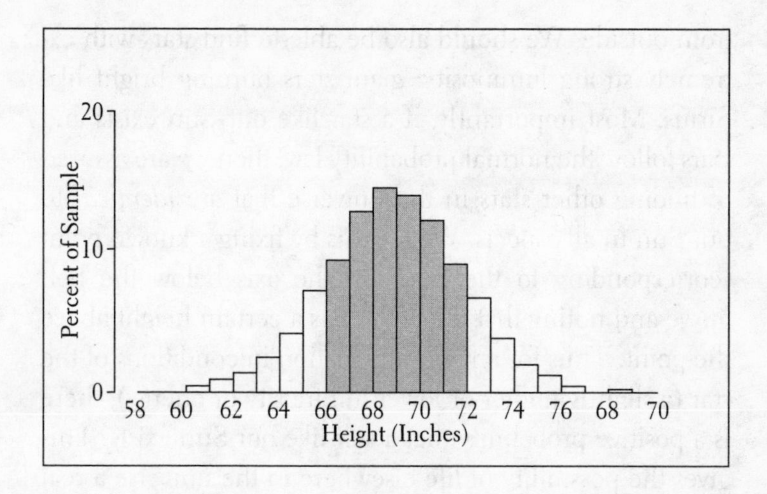

extend all the way to minus infinity on the left and plus infinity on the right, so that there is an infinitesimally small probability of falling at great distances away from the average. This means that, theoretically, one can find a person with a negative intelligence, as well as one with an IQ of 300, although the probabilities of these two events are extremely small—but not zero.

The distribution of the luminosity of stars follows a normal distribution, as do the distributions of many other properties of stars: color, temperature, size, and so on. In terms of its luminosity, our Sun falls in a region that contains an area of 5 percent under the normal curve, so there is a 5 percent probability that a randomly chosen star will have the luminosity of our Sun. The normal curve can thus be used to obtain probabilities of various aspects of a star. According to the normal law, we should be able to find extremely dark stars, with luminosity of zero or even negative: black holes that actually swallow up radiation from outside. We should also be able to find stars with extremely strong luminosity: giant stars burning bright like Sirius. Most importantly, if a star like our Sun exists and stars follow the normal probability law, then we are assured of finding other stars in the universe that are *identical* to our Sun in all aspects. We see this by fixing a known point (corresponding to the Sun) on the axis below the bell curve and noting that the curve has a certain height above the point. Thus for any slight variation in conditions of the star (a slightly higher or lower luminosity or energy), there is a positive probability that a star like our Sun exists. This gives the possibility of life elsewhere in the universe a real probability, in that a Sunlike star giving energy to an Earth-like planet must indeed exist.

In physics, however, the most useful applications of the normal curve occur in the quantum theory. The reason is that the normal curve, in its very essence, is a limit theorem for a very large (infinite) number of random factors. In most applications, the number of factors—say, the number of times a coin is tossed—is finite and not even very large. In statistical mechanics, physicists study the collective behavior of an immense number of atoms or molecules or electrons. Here, because the number of elements under study is so huge, the normal curve serves not only as a good approximation as it does in other fields, it works as the *exact* probability rule. The normal curve is thus indispensable in quantum physics, for it yields the exact results the physicist needs.

Albert Einstein became uncomfortable with quantum theory. It dealt with electrons, not stars, and his own work dealing with electrons—his research on the photoelectric effect, which won him the Nobel Prize—did not lead him to probability. It also seems that Einstein was opposed to the idea of probability in describing the world. Consequently, faced with the probabilistic aspects inherent in the quantum theory, Einstein made his famous statement: "I shall never believe that God plays dice with the world!"

In 1892 the French mathematician Henri Poincaré discovered that some mechanical systems—such as pendulums, whose laws of motion were known and modeled well by the equations of physics—sometimes displayed unexplainably chaotic behavior. Poincaré also studied the famous three-body problem in physics. Here, he obtained a solution that baffled him with its complexity. He described the relationship between two curves in his solution, whose intersections formed a complicated web where each curve folded back on itself infinitely many times. Unbeknownst to him, Poincaré was looking at one of the most bizarre creatures of mathematics: something we call today a strange attractor. Poincaré gave no further thought to the curves he was studying—they were too complex to be of use—and gave up on trying to understand what he had inadvertently discovered. Today we know that Poincaré's solution was a manifestation of chaos.

Almost seventy years passed, and in 1963 the meteorologist E. N. Lorenz of the Massachusetts Institute of Technology made a discovery that would change forever how mathematicians, physicists, and other scientists viewed chaotic behavior. Lorenz was studying weather systems and he modeled them using three coupled equations. The equations were nonlinear, meaning that their graph was not a straight line. Nonlinear equations often lead to chaos, and in fact chaos is a particular kind of nonlinearity in nature. In a linear equation, a change in one variable will result in a proportional change in another variable. If a commodity costs $5 then buying 10 units will cost you $50. When nonlinearity is present, the effects on a variable are of a different, nonproportional magnitude. The simplest case would be if total price was obtained as $5 times the square of the quantity bought (so 10 units would now cost $500). Chaos is an extreme form of nonlinearity in the sense that not only is the output not proportional to the input, but it is also unpredictable.

Lorenz solved his complicated nonlinear equations for the weather systems and plotted them. What he got was a strange attractor: a region to which all sufficiently close trajectories are attracted in the limit, but in which arbitrarily close points become exponentially separated in time. The Lorenz strange attractor is shown below.

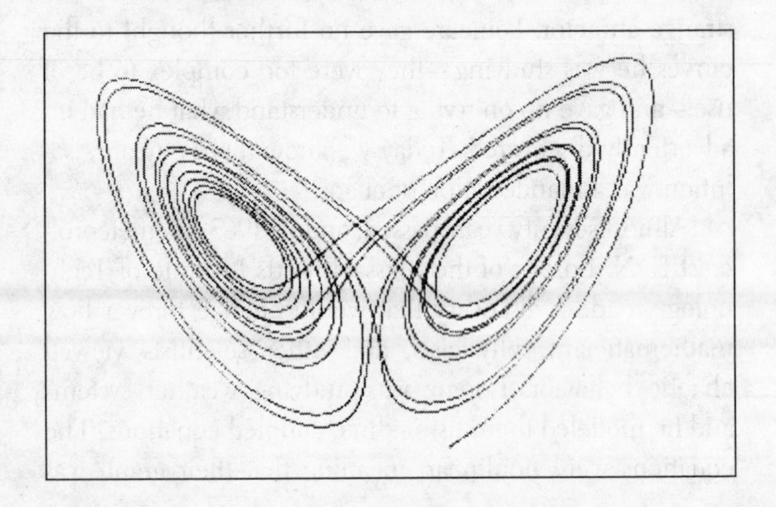

The strange attractor Lorenz found for weather systems is characterized as exhibiting the butterfly effect. Since the Lorenz system is so sensitive to initial conditions, it is said that a butterfly's stirring of the air in China could cause the slight shift in initial conditions that would create a hurricane off the U.S. East Coast a week later. Thus was born the mathematical theory of chaos—a field of science that would give us a whole new way of thinking about the world and would shed new light on Einstein's challenge to science.

The word *chaos* is derived from the Greek, where it originally meant the infinite empty space that existed before creation. In everyday language, the word conjures up images of people running every which way or items flying in the air in no apparent order. It is almost synonymous with *randomness*. But the mathematical theory of chaos couldn't be further from randomness. A chaotic system is the exact opposite of a random system. A chaotic system is perfectly *deterministic*—it has absolutely no random element to it. It only looks random, but everything about it is precisely specified by a mathematical rule, an equation. In chaos there are no dice. And, therefore, if every inexplicably chaotic behavior in the world could be described by a mathematical equation, then Einstein would be right about God and dice.

The best example of a chaotic system is the baker's transformation. This is a simple-looking repetitive transformation of numbers to form a continuing sequence. It is also called the logistic mapping and is given by: $x \rightarrow kx(1-x)$. Here x is a number between 0 and 1, the variable, and k is a fixed quantity, say $k = 3$. Each number is obtained from the previous one by plugging it into the equation. Let's start with $x = 0.2$. Using $k = 3$, we get: $3(0.2)(0.8) = 0.48$. Now the number 0.48 will lead to the next number in the sequence: $3(0.48)(0.52) = 0.7488$. Continuing this way, we obtain a large sequence of numbers. This system is chaotic. This means that it is extremely sensitive to initial conditions. If we start the sequence with a number just slightly different from 0.2 as we started this time, the results would be very different.

The logistic mapping "folds" the line segment from 0 to 1, in the sense that numbers between 0 and 0.5 are mapped by the transformation to numbers between 0 and 0.75. Numbers between 0.5 and 1 are mapped by the transformation to the same interval 0 to 0.75, but in reverse order. Thus, every time a new number is derived from the previous one, a "folding" of the interval occurs. This folding of the line is exactly like the kneading of dough in the hands of a baker. Suppose that two raisins are inserted into the dough right next to each other. The baker goes through the usual kneading, stretching, and folding the dough, not looking to see where the raisins are. After a while—no matter how close the two raisins were to each other when the work began—there is no way of predicting where the two raisins will be at the end when the dough is ready to go into the oven. This is a manifestation of chaos. For even a small difference in the initial placement of the two raisins can translate into a big difference in the final position of the two raisins.

Chaotic systems appear in many places in nature. One example is a driven pendulum. As you push a pendulum, it swings faster. You push it again and it swings faster yet.

Then, after a few more pushes, there comes a moment when the pendulum goes wild and swings in an erratic way. This is chaos. The pendulum's position and velocity can be drawn on a graph with one direction being the position, the other the velocity. This is called a phase diagram. In the phase diagram, one would notice from the solution of the equation governing the pendulum's motion that there is extreme sensitivity to initial conditions. This is just like the sensitivity of the particular weather system studied by Lorenz where something as small as a butterfly can cause big changes in the weather far away.

There are many other physical systems that exhibit chaotic behavior. Some chemical reactions, too, are chaotic in their nature. The reaction time and extent depend in a very strong way on the initial conditions. For example, a chemical reaction will occur smoothly if the conditions are right, but then a slight—even undetectable—change in the conditions may change the reaction to an explosion.

Stephen Smale, who derived the important differential equations of the predator-prey relationship, did extensive work on nonlinear equations and chaos. In 1967 he published a seminal paper that explained much of what we know about chaos and strange attractors. Years later I went to see the movie *Jurassic Park.* It was clear that the mathematician played by Jeff Goldblum was none other than Stephen Smale: both wore the same black leather jacket, had the same distinct speech pattern, and were concerned with strange attractors. In the book, Michael Crichton went further than in the movie about the idea of chaos. When the dinosaurs were cloned, the scientists lacked one small piece of DNA and they got it from a frog. This was the small change in initial conditions that led to the unexpected final outcome, as happens with all chaotic systems. Here, the result was that the cloned dinosaurs ran amok. But could this happen in nature, or only in fiction?

The question of whether life systems, and other systems in the universe, are random or chaotic (and thus deterministic) is probably one of the great unanswered questions in all of science. There is no doubt that chaos and extreme nonlinearity exist in biology, physics, and other areas. It is possible that certain mental illnesses, for example, manifest themselves when a small initial change in the level of a hormone in the brain takes place. What concerns us here is the question of whether the development of life on Earth occurred as the result of a random fluctuation in the primordial soup of chemicals present on the surface of the early earth, leading to the creation of DNA, or whether DNA and life were created by a nonrandom, deterministic process governed by a nonlinear equation. Was the emergence of life on Earth the result of chance or choice (the latter being a complex, nonlinear prescription)? This deep question holds the key to the possibility of life elsewhere in the universe.

One perplexing property that characterizes the strange attractors, these graphs in phase space of chaotic systems, is that the graphs do not fill the entire space in which they are embedded. Imagine smoke filling an enclosed space such as a small room. As the smoke dissipates into the air, it will eventually fill the entire three-dimensional space of the room. The reason for this is thermodynamics: the ever-present law of increasing entropy. Since there are many more ways in which a tremendous number of smoke particles can arrange themselves in the *entire* room, they do so, rather than restrict themselves to a subspace within the room. It is much less likely—the probability for it is fantastically small—that the smoke particles will stay only in one part of the room, given enough time. This is a probability argument from the area of statistical mechanics, where the normal curve reigns supreme. Not so in chaotic systems. Here, while the phenomenon may *seem* random in a cursory inspection, the system is very well determined. The smoke particles, to continue the analogy, will not fill the entire room when they are governed by a chaotic nonlinear system of equations—they will fill some smaller subspace of the room, with dimension less than the usual three dimensions. Look at the figure showing the Lorenz attractor. While the entire space is three-dimensional, the attractor does not fill the entire space; it looks like the tilted wings of the butterfly in the analogy of the chaotic weather patterns it supposedly describes. The attractor is two-dimensional, but it is folded at an angle into the third dimension. We say that the attractor's Hausdorff dimension (named after a German mathematician), which represents the actual dimensionality of the surface, is between 2 and 3—it is equal to $D = 2.06$.

While all the work on chaos and strange attractors was going on, a French mathematician who came to the United States and to work at IBM labs and Yale University made a breakthrough discovery, which on the surface seemed to have nothing in common with chaos. Benoit Mandelbrot is the last in the line of the great French mathematicians who are experts in a wide variety of fields, in the tradition of Fermat and Poincaré. In the early 1960s Mandelbrot made important discoveries that allowed financial analysts to understand the movements of stocks. He did pioneering work on coding theory, statistical thermodynamics, linguistic statistics, communication theory, and random walks. In 1974 Mandelbrot turned his attentions to processes of life and nature. He first approached these topics from a probabilistic and statistical point of view, assuming an inherent random component to all of life's variables. He applied his methods to the number of mutants in old colonies of bacteria. Later, Mandelbrot looked at other natural processes, such as geological structures and the craters of the Moon. What he found was far from random. To describe natural processes, Mandelbrot derived a new concept. He called it a fractal.

Mandelbrot posed this question: How can one measure the length of the coast of Britain? One could start somewhere on the coast, Mandelbrot reasoned, and pull a rope from a post along a bay in a straight line to another post, then continue until one has exhausted the entire coast of Great Britain. But would this measurement be accurate? Wouldn't it miss the curvature of the bays and inlets that were crossed by the straight lines one was drawing? A better way, he reasoned, would be to place the posts closer together so that the curvature of bays and inlets could be better accounted for in the total measure. But looking at the rugged British coast, Mandelbrot realized that the measurement should be done even more finely, to account for large rocks by the sea, with their jagged edges. This would complicate things further since the ropes would have to follow the rocks and not only the bays. But looking closer yet, he realized that there are objects on the coast of Britain that are even smaller than the jagged rocks but of the same general form. These objects, the little stones by the water, would also have to be accounted for. A finer measuring method would have to be designed. And once this was done, there were even smaller objects than the stones, with the same general shape as the stones and rocks and little hills, and so on to grains of sand with the same jagged outline as the rocks and the bays and the entire coastline. Mandelbrot realized that his argument could be continued indefinitely: on the coast of Britain there are smaller and smaller objects that are self-similar: they have the same general shape as the larger objects. How would one measure the entire length of rope needed to curve around the smaller and smaller objects?

Mandelbrot realized that the dimension of the area-filling curve that was made of finer and finer turns around smaller and smaller objects was somewhere between that of a line and that of a plane. He was searching for a definition of this measure and came to the idea of a fractional dimension: a dimension that is not an integer such as 1 (for a line) or 2 (for a plane) or 3 (for three-dimensional space). Mandelbrot named the fractional dimension he discovered a fractal. The coast of Great Britain, Mandelbrot concluded, is a fractal: an object that is a plane-filling curve whose shape is self-similar on any scale one uses. It is something with dimension between 1 and 2, since it is not quite a plane, but no longer a line: it is infinitely curved around itself.

Mandelbrot understood that what he discovered in nature in the form of the coast of Great Britain was also something that existed as a purely mathematical construction. As a mathematician with knowledge in a wide variety of areas, Mandelbrot was well aware of the Cantor set. This was a strange mathematical object discovered by Georg Cantor (1845–1918), a Danish-German mathematician who spent his life trying to understand the idea of infinity and derived a whole theory to describe this difficult idea, only to end his life in an insane asylum in the German city of Halle at the end of World War I. Cantor defined his set as follows: Start with the interval from 0 to 1. Now remove the middle third of the interval, leaving the two remaining thirds on each side of it. For each of the two remaining intervals, remove the middle third. Then continue in this way, always removing the middle third of every remaining small interval. In the limit, once this has been done an infinite number of times, we get the Cantor set.

The Cantor set lives on a straight line, but its measure is less than that of the line, since it has infinitely many large holes in it. Its measure is, therefore, less than 1. Mathematically, we define the Hausdorff dimension of the Cantor set as 0.63 (it is the log of 2 divided by the log of 3). The Cantor set is a fractal. It lives in a space less than that of a line. There is an interesting fractal that lives in the space between that of a line and that of a plane. It is called the Koch curve. Start with an equilateral triangle. Divide each side into three equal parts and on the middle part build a new equilateral triangle. Then continue this operation infinitely. The Koch curve is shown below.

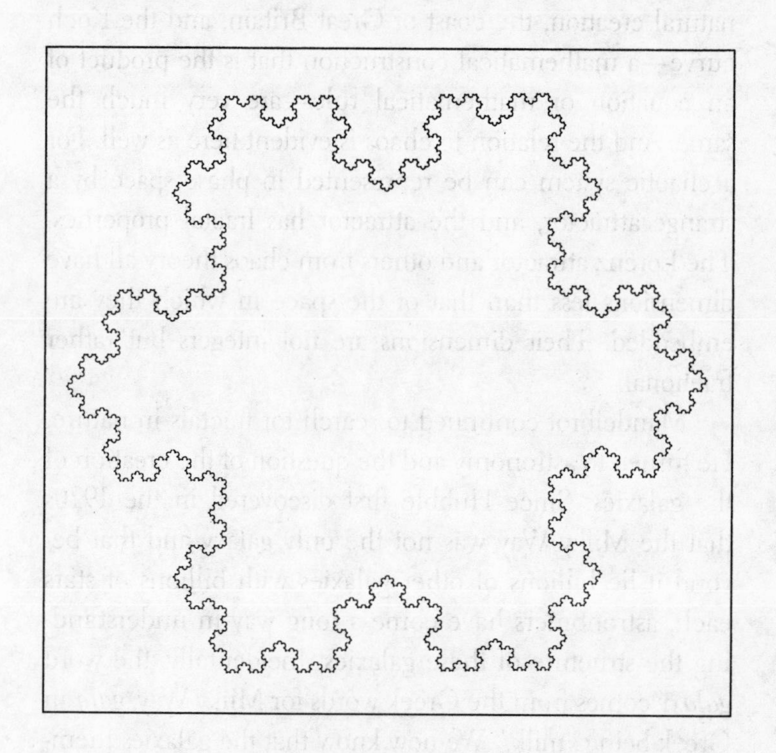

The Hausdorff dimension of the Koch curve should be between that of a line and that of a plane because the curve is a line that has infinitely many kinks so it almost fills a planar area, but not quite. The dimension is 1.26 (log 4 divided by log 3). The Koch curve resembles the coast of Great Britain. It is jagged in a way that repeats itself on smaller and smaller scales ad infinitum. Look at the edge of the curve through a microscope and you will see the same pattern you see with the naked eye. Increase the power of the microscope higher and higher and you always see the same pattern. The Koch curve exhibits the characteristic self-repeating property of a fractal. Thus Mandelbrot's view of a natural creation, the coast of Great Britain, and the Koch curve—a mathematical construction that is the product of an equation or mathematical rule—are very much the same. And the relation to chaos is evident here as well. For a chaotic system can be represented in phase space by a strange attractor, and the attractor has fractal properties. The Lorenz attractor and others from chaos theory all have dimensions less than that of the space in which they are embedded. Their dimensions are not integers but rather fractional.

Mandelbrot continued to search for fractals in nature. He turned to astronomy and the question of the creation of the galaxies. Since Hubble first discovered in the 1920s that the Milky Way was not the only galaxy and that beyond it lie billions of other galaxies with billions of stars each, astronomers have come a long way in understanding the structure of these galaxies. Incidentally, the word *galaxy* comes from the Greek words for Milky Way, *gala* in Greek being "milk." We now know that the galaxies themselves are arranged into groups. There is a local group of galaxies, a cluster, and the clusters are grouped together in superclusters, and so on. What was surprising for astronomers was to find that the galaxies and their clusters are not randomly scattered throughout the known universe. There seemed to be a strange order to their arrangement. But what was that order?

Mandelbrot looked at the galaxies from several different hypothetical vantage points. The first one was that of Earth. As seen from Earth, the galaxies clustered in a way that filled the celestial plane in a pattern. He then constructed a view of the galaxies from the vantage point of the Centaurus constellation. Mandelbrot noticed that the galaxies and their clusters exhibited the same space-filling pattern as they did from the Earth viewpoint. He then changed the focus and looked at galaxies that are within several thousand light-years of Earth and the same pattern appeared. Mandelbrot concluded that the structure of the galactic filling of the universe, as projected onto a plane, was a fractal. He computed its dimension as 1.2.

Mandelbrot and others have since discovered fractals in many places in nature, from leaves on trees to weather turbulence. There has even been a suggestion that fractal patterns may be embedded in thought processes in the human brain. Chaos and fractals and strange attractors are all manifestations of structure. The appearance of such phenomena in nature seems to want to tell us that there are rules, some of them very complicated, that govern life and the universe. The question that arises is whether life itself began on Earth as a result of a mathematical prescription or a random event or sequence of events.[16] Does God play dice or command with equations?

Did the very structured DNA molecule come about by a chance sequence of chemical reactions, or was it formed by a specified rule, like a fractal out of the chaos of creation? Whatever happened here on Earth should hold the key to the mystery of whether life exists elsewhere in the universe. And another question comes to mind as we consider the possibility of a chance creation versus that of some hidden well-determined mechanism: What is the reason that all the amino acids involved in processes of life on Earth are of the L-form? If chance alone were at play here, we should expect that half of them would be of the L-form and half would be of the D-form. Both molecules are identical in terms of structure and reactivity and every other parameter describing them. It is only the symmetry that separates them. If all of these important molecules found in the processes of life are of only one kind, is there a mysterious deterministic force that made them this way? This fact may be a strong argument against random fluctuations and for determinism. This is not to say that we do not see chance and randomness at play everywhere on Earth. It seems that both chance and deterministic factors are inherent in the processes of nature. Randomness is clearly a force to be reckoned with.

But what is randomness? Is it indeed separate from the realm of mathematical equations—simple, nonlinear, or chaotic? Randomness is often associated with games of chance. A die is cast, and we believe we see in its roll a manifestation of pure chance. It rolls on a table and stops somewhere no one can predict. The number the die shows is random; it is due to chance. But if we had infinite wisdom and we could actually see how the die is falling out of our hand, measure its velocity and angular momentum with infinite precision, observe it hit the surface of the table on one of its edges, measure its angle of impact—then use all of these parameters as well as the air pressure and minute currents and vortices of air produced by the moving die. We should thus be able to compute exactly the number that will appear on the die.

For randomness is randomness only because of our ignorance. If we knew the values of all the physical variables involved in the motion of an object such as a die, we should be able to construct the object's equations of motion, solve them (not always an easy task), and know the exact outcome. When the die hits the table on one of its edges, we have a manifestation of chaos. This is true because two different rolls of the die where the die hits the table at *almost* exactly the same momentum and angle will not produce the same result. The corner of the die hitting the surface exhibits great sensitivity to initial conditions. A throw of the die may result in the number six and the next throw—everything being identical except that the angle the edge of the die makes with the surface of the table is 0.00001 degrees greater—may result in a different number. But if we knew everything *precisely*, we could predict the outcome. So randomness and chaos are not really two very different concepts. The difference between them is information. And in fact all of the theory of probability depends on information. Probabilities are measures of our knowledge or ignorance about the world. Probabilities are conditional on our set of information.

When I teach probability theory, I often ask my students the following question not long after introducing the concept of probability: "What do you think is the chance that the stock of a small computer company will go up in price tomorrow?" Usually the students hesitate since they know next to nothing about the company. But when I explain that probability is valid in various contexts and that they can give me their "personal" probability for the event, assessed in some logical context, I usually get the answer: "Well, maybe 60 percent. . . . The market has been going up overall, and computer companies have been doing pretty well . . ."—which is a good answer. Then I ask them: "Suppose that you have a piece of information nobody else has. You know that Bill Gates and Microsoft have just put in a bid to buy this small computer company. The announcement of a possible merger will be made in the morning. What is *now* the chance the stock will go up?" The answer I get is "Close to 100 percent," correctly reflecting the fact that companies about to be bought up by bigger ones almost always enjoy a surge in stock price once the announcement of a merger becomes public. The moral here is that probability depends heavily on information. There is a reason illegal insider traders go to jail—the game is tilted unfairly in their favor. They have information that makes the market nonrandom within their *own* information set. So randomness or determinism depend wholly on what you know or don't know. Chance and probability serve as a good model when there is inherent uncertainty, when we don't know things precisely.

A computer cannot do anything random. Computers, by their very nature, do exactly what you tell them to do, and it's impossible to tell the computer to do something "random"—it doesn't know how. Perhaps this is one key difference between human intelligence and that of the computer. Human beings sometimes, for no good reason, will do something wild just for the fun of it. You'll never find a computer given to whimsical behavior. The only way to have a computer generate a truly random number is to make it command an artificial arm to roll a die and note the outcome. Since this is such an inefficient way of doing things—especially as we use computers for their tremendous speed—you will never see this done anywhere. But computers do generate random numbers, don't they? The entire theory of simulation, which requires the use of extensive randomness, is implemented on computers. And computer games almost always have at least some random component to them. So how is this accomplished?

The answer is that computers generate pseudorandom numbers—not real random numbers. This is done in a precisely specified way since the computer must be told exactly what to do and how to do it. The pseudorandom numbers result from a deterministic program within the computer (and even some advanced calculators offer this capability) that uses a chaotic algorithm. The computer starts with a seed—a single, prespecified number. Then it carries out an arithmetic operation on the seed that is just like the rule that generates the sequence of the logistic mapping, the baker's transformation. A new number is obtained and it is taken to be "random." Then from this number, through the same transformation, another number is obtained, and so on. All the numbers are precisely well determined by the equation of the transformation, but—as everywhere in chaos—these numbers *look* random. The problem is that if a new sequence of "random" numbers is required, the computer will give you the same sequence, unless you change the seed. So in all computer operations, chaos serves the purpose of pretending randomness. Again, the two concepts are intricately intertwined.

So where does the line pass between pure randomness and pure determinism? Statisticians have a test that can tell them when numbers or symbols are the result of randomness or whether there is some pattern to them. The test is called the runs test. You enter a bar and there are twelve bar stools and six people sitting at the bar. Using the letters O for occupied and E for empty, this is the pattern you notice in the way the six people are sitting at the bar: O E O E O E O E O E O E. Is this pattern due to chance, or did the people purposely seat themselves away from others? Common sense tells us that the answer is the latter. But statisticians have a way of quantifying what we sense is the right answer here. A run is an uninterrupted sequence of symbols of the same kind. Here there are twelve runs (which is the maximum number of runs of twelve objects, six of each kind), so this sequence is most probably not random. Now consider the other extreme. You walk into the bar and you find people occupying the seats in this order: O O O O O O E E E E E E. Of course common sense tells us that these are friends all sitting together. They didn't choose their six seats at random out of the twelve available ones. Statisticians count here two runs (a run of six Os and a run of six Es). This is the smallest number of runs of twelve objects of two kinds, with six of each kind. This, too, with high confidence, is not a random selection. But what about the case O E E O O E O E E O O E. Here, the answer is not clear. The number of runs is eight, which we say is not statistically significant. There is no strong proof that these seats were *not* chosen at random. A statistical table can give us the range of values for the number of runs that will lead us to strong proof of nonrandomness.[17] The runs test can be used when the raw data are numbers rather than symbols. Is the following sequence of numbers random or not: 5 3 1 7 3 7 8 4 2 4 6 6. To answer this question, denote the numbers as O for odd and E for even. Then the sequence above is O O O O O O E E E E E E, and we have the same situation as the bar stool example with two runs. This is most likely not a random sequence of numbers (even though it looks random!).

When pseudorandom numbers produced by a computer are subjected to the runs test, they pass the test in that there is no evidence of nonrandomness. So the chaotic mechanism in the computer, which precisely determines the numbers to be chosen, with no real random element in this choice, acts for all practical purposes just like a die or a roulette wheel or a deck of cards. Randomness and extreme nonlinearity are very closely—often indistinguishably—related to each other.

What does this tell us about DNA and its appearance on Earth 3.5 billion years ago and about the prospects for this amazingly complex molecule of life to have appeared elsewhere in the vast universe? Is the creation of DNA random or deterministic? Chance and predictability are really two ways of looking at the same thing. Is the insider trader playing dice or acting on the outcome of some equation? Are pseudorandom numbers random or do they form a sequence with an exact but unknown structure? The answer depends on your point of view. An equation represents complete knowledge (this assumes that the equation captures all the variables that play a part in the phenomenon in question). Chance reflects at least some ignorance about a phenomenon. There is an area of applied mathematics called stochastic processes. Here, we use equations that *also* have a random component in them. This is an attempt to deal with life both ways: the equation captures much of what is known about a process, while its random component captures the essence of everything that we still don't know about the situation. The known parameters are plugged into the equation, and for the random components that represent the fuzziness in our knowledge of the world, known averages and standard deviations are plugged in. Such hybrid models of perfect knowledge with uncertainty have worked very well in science.

The lesson here is that we must use everything that we know with precision, and use probability theory to handle the unknowns of life. Probability theory should, therefore, provide a good tool in trying to evaluate the prospects for the existence of life elsewhere in the universe. Most likely, DNA was created as the outcome of a complicated system of nonlinear equations that govern the rates of chemical reactions in the presence of catalysts in a rich solution of methane gas, water, carbon dioxide, phosphorus, nitrogen, and other compounds with the right temperature, time, and other conditions. Knowing the exact set of complicated equations and how they react chaotically would tell us if life exists on a given extrasolar planet once we know the exact conditions there. But unfortunately, we do not have this knowledge. We can, however, approach the problem using the theory of probability. This is our powerful tool for handling uncertainty in all situations. If we make the correct assumptions and use them well in constructing a probability model, we should come up with a good answer. Knowing that a company is about to be bought by a larger one gives us a high probability that the firm's stock will go up. This is an example of a reasonable assumption within a probability framework.

Statistical arguments also work well in building a probability model. If we have the statistical distribution of people's heights, for example, then while we may not *know* the height of a randomly chosen individual (hence we are in a state of ignorance), we can still compute the *probability* that the individual's height will fall within any range of values from the statistical distribution we do possess. This is how knowledge (about the entire population) can be used to estimate something about which we have no knowledge (the individual). If out of nine newly discovered extrasolar planets, one is known to orbit its star within the habitable zone—the zone where water stays in liquid form—then the chance that any planet falls in the habitable zone can be estimated as 1/9. Here, we don't know the statistical distribution of all the planets in the universe, but the knowledge we do possess from our very limited sample should still give us a rough estimate of this parameter.

FOR DECADES PHYSICISTS have tried to explore whether chaos plays a role within the quantum theory. Are there small quantum "pendulums" that swing and enter a chaotic mode when the driving force reaches a certain level, as happens in the larger nonquantum world? Some results have indicated that a form of chaos may exist in the microworld of quantum mechanics. But the nature of this chaos is very weak. There is no dramatic jump into chaos as one finds in the macroworld of large systems. In the quantum world, the wave function seems to reign supreme and to account for most of the motion exhibited by small particles. Here, it seems, there is no escaping the strong grip of probability theory and of randomness and chance. The solutions to quantum equations are probability distributions—not exact statements—although this may simply mean that we can't escape our own ignorance about quantum systems. The position of an electron at a given point in time can still not be described exactly—all we have is a probability distribution. At least within the quantum world, God does play dice. But He always knows the outcome.

9 The Inspection Paradox

HAVING DECIDED THAT, whether the universe is random or deterministic, or both, probability theory may be used to model phenomena about which we have little or no knowledge, we are in a position to use probability models in considering the problem of the existence of life elsewhere in the universe. Probability is intimately related to statistics—the science of information.

In today's world, people are bombarded daily with statistics. Television, radio, newspapers, and magazines all give us numbers until we are numb to them and have no feeling for what information these numbers may convey. Sometimes statistics are misleading, where one economic indicator seems to imply growth and another implies stagnation. And people we consider "experts" often have no idea how to interpret the numbers. This doesn't mean they don't give us their views anyway. Because numbers in today's economy are so large, people often lose sight of the magnitudes of such numbers. When we hear that some individual is worth $50 billion, the word *billion* somehow becomes smaller in our eyes than the immense size it has in reality. For a billion is a very, very large number. And the general public's grasp of astronomical concepts suffers because of this inflation of numbers in the news. Our galaxy has hundreds of billions of stars. That is many, many stars. We couldn't count them all, even if we were able to see each and every one of them. If it takes us a second to count each star, counting continuously without rest, it would take us ten thousand years to count the stars in the Milky Way galaxy.

And the Milky Way is only one galaxy out of many billions of galaxies in the known universe. The number of stars in the universe is immense beyond anyone's imagination. And the fact that the federal deficit was so large that it was measured in such numbers should in no way diminish the sheer size of the universe in anyone's mind. When searching for life in outer space, we must remember that there are many, many places to look. And the distances among stars are as mind-boggling as the number of stars. The nearest star, Proxima Centauri, is 25 trillion miles away. Flying at 30,000 miles an hour would take a spaceship almost 10,000 years to reach this nearest of all stars.

But number inflation is not the only problem with the public's understanding of statistics. Lying with statistics is something that occurs every day in the media and elsewhere. The lying is not always intentional. We often misunderstand what the numbers are trying to tell us. The most common form of "lying with statistics" is the misrepresentation of quantities on graphs. In many cases there is no clear way of deciding what should be the scale of the two axes on a graph.

In order to understand numbers and statistics and try to make sense out of them, it is important to understand two measures: the mean and the standard deviation. The mean is a measure of centrality of a statistical distribution: it is the average of the population of interest. If the distribution is that of probabilities rather than a statistical frequency distribution of a population, the mean represents the expected value of the random quantity of interest: it is the value that would be the average of a large number of random realizations from the probability distribution. Using the stock market as an example, the expected value is what you would *expect* the stock market to yield. It is the average yield over many potential realizations of today's market conditions. For the normal (Gaussian) distribution we saw in the last chapter, the mean of the distribution is the number lying right in the center of the bell-shaped curve.

The standard deviation of a distribution is a measure of the spread of the statistical or probability distribution. The wider the curve, the more variability, or uncertainty, is present in the possible values of the random quantity of interest. In a stock market context, a riskier market is one with a wider curve of possible returns.

The mean, the expected value of a random quantity, is the most important statistic. But the mean can be very deceiving in many situations. For one thing, the mean is very sensitive to outlying observations (called outliers). The following are luminosity data on a sample of ten stars: 2.3, 2.5, 2.1, 3.0, 2.8, 2.9, 2.8, 2.6, 2.3, 7.1. The mean, the average luminosity, here is 3.04. But the average has been affected strongly by the outlier 7.1—an unusual observation that is very different from the rest of the stars in the sample.

Without it, the mean would be 2.59, which is more representative of the data set as a whole without the outlier. It is situations like these that bring about the need to use an alternative statistic, the median. The median is an observation that would lie in the middle of the data set; here it would be 2.7. The median is not sensitive to outliers.

An amazing phenomenon that can arise from the use of the mean is the inspection paradox. Murphy's Law says that if something can go wrong, it will. This pessimistic maxim about life in general probably came out of some real observations about aspects of everyday life. Suppose that you live in a city and take the bus to work every day. You know from the bus schedule that, on average, a bus arrives at your stop every ten minutes. You reason, therefore, that if you go down to the bus stop at a random moment, you would have to wait—on the average—about five minutes. This is so because ten minutes is the total average time between buses, so on average you should arrive at the stop about halfway between the two buses, and therefore your average wait should be five minutes. But from your experience, you *know* that this is not true. It seems you always wait longer. Your average wait is longer therefore than what by all rights it should be: five minutes. Why does this happen? Is Murphy's pessimism always justified?

The fact that you can expect to wait longer than the average wait is absolutely true. This is a manifestation of the inspection paradox, which can be proven mathematically. But Mr. Murphy's looking at the bad side of everything may not be justified. For the inspection paradox often works in your favor as well. Suppose you have a flashlight with a battery.

You know from the manufacturer's claim that the average life of this battery is 3.5 hours. But you know that you've used your flashlight with this same battery inside it for much longer than 3.5 hours, and it's still good. The battery in your flashlight right now will last longer than the average battery. This, too, is a true fact. It is another manifestation of the inspection paradox. And it can be proved mathematically with absolutely no ambiguity. You wait longer than average when you go down to the bus stop. But your flashlight battery also lasts longer than the average battery.

Trying to make sense out of the inspection paradox is an exercise in understanding probability. Doing so will also shed light on the problem of the existence of extraterrestrial life. Let's suppose that you are now going down to the bus stop. By the inspection paradox, you will likely wait more than the average wait for a bus (the old Murphy's Law and the persistence of bad luck). Let's see why this is true. Buses arrive randomly at your bus stop, but the average interarrival time is ten minutes. This means that sometimes the bus comes in less than ten minutes after the last bus departed, and sometimes more than ten minutes after the departure of the previous bus. The picture below shows some typical bus interarrival time intervals. An *x* marks a

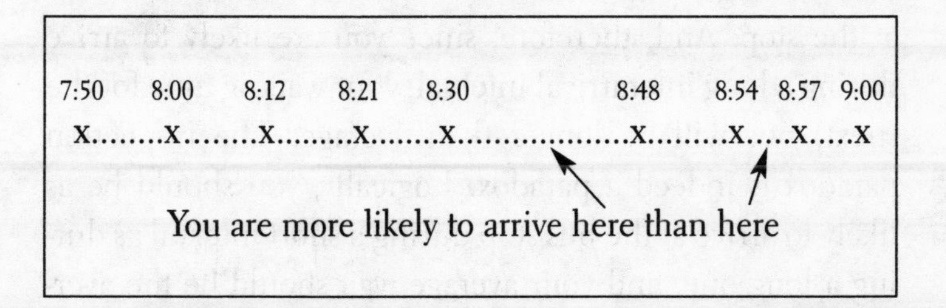

bus's arrival. Notice that some intervals are long (longer than the average ten minutes) and some are short (less than the average ten minutes).

Now think of your arrival time at the bus stop as a dart randomly thrown at a dartboard, except that the dartboard here is one-dimensional: it is the straight line of time, shown above. You arrive at a random point on the time line. This is a crucial point, since random events are governed by the laws of probability. And what do the laws of probability say about your arrival instant on the time line? They say that your arrival time will have a higher probability of occurring during a long interval than it has for your arrival during a short interval. Why? Simply because a long interval of time gives you more possibilities of arriving during an instant it contains than a short interval. You are much more likely to receive a phone call during an interval of an hour than you are during an interval of one minute. This assumes equal likelihood for each subinterval of time. If someone will call you randomly sometime in the evening, you are more likely to receive the call if you wait for it between 7 and 8, than between, say, 7:23 and 7:24.

Let's return to the waiting for the bus problem. When you go down to the bus stop, you are most likely to arrive there during a long interval between two arrivals of the bus at the stop. And, therefore, since you are likely to arrive during a long interarrival interval, your waiting time for the (next) bus will be longer than average. The inspection paradox is indeed a paradox. Logically, you should be as likely to arrive at the bus stop during a short interval as during a long one, and your average wait should be the average half interarrival interval of five minutes. But since you are more likely to arrive randomly during the longer inter-arrival intervals when you go down to the bus stop, your wait is expected to be longer than the average wait!

The same argument works for the good side of the paradox: your present battery lasting longer, on average, than the average battery. When you pick up your flashlight, you are choosing a random point in time, and the changing times of batteries in your flashlight are like the arrival times of the buses. Picking up the flashlight is like arriving at the bus stop. Since it is more likely that this random event will occur during a long period between your batteries dying than during a short period, the present battery in your flashlight has a longer life than the average battery.

What's behind the inspection paradox is the fact that distributions that have already begun their lifetimes have longer than average life spans. If you arrive at the bus stop, watch the next bus arrive and leave, and *then* wait for the following bus, then you can expect to wait the average interarrival time of buses. Here the inspection paradox no longer applies since the distribution of bus interarrivals is starting at the beginning (the bus that had just left). You are not arriving during a distribution that has already started its life, as in the case of arriving after the last bus had already departed. The fact that distributions that have already begun their lifetimes have longer lives than average (the inspection paradox) has implications for longevity studies. A living person has a longer expected lifetime than someone not yet born. A thirty-year-old man can't die from a childhood illness and in fact can't die at any age younger than thirty—he is therefore likely to live longer than someone just born, who can still die young. An eighty-six-year-old woman can no longer die at any age younger than eighty-six. She therefore has a higher expected (that is, average) age at death than that of the average woman.

The inspection paradox recently tripped up the Israel Bureau of Statistics. World longevity data came out in 1994 indicating that, for men, Japan was the leader in longevity, while in second place was the state of Israel, where men live an average of 75.1 years. The entire country was celebrating the good news, no one suspecting that there was a little problem with the interpretation of statistics. In a land where a large proportion of the population has immigrated from outside, the longevity figure was deceptive. Since a living person has a longer expected total lifetime than one not yet born, and since many old people immigrate to Israel, even further increasing the expected life of people living there, the statistic was highly biased and overestimated the actual longevity of people born in the country.

WHAT DOES THE INSPECTION PARADOX have to do with life elsewhere in the universe? Let's return to the bus stop example. You choose a random time and walk down to the bus stop. Because you are much more likely to arrive at the stop during a long interval between the arrivals of two buses, your expected wait at the bus stop when you go there now is longer than the average wait at the bus stop. When randomly appearing at the bus stop, you likely "catch" a long interval. The longer a bus interarrival interval, the more likely you are to fall within it.

Now let's assume that there are other living cultures out in space, some advanced, others not; some intelligent, others primitive. Some of these collections of extraterrestrial beings have been around for only 100 million years and have evolved into simple life-forms. Others have existed for 4 billion years and have evolved into humanlike civilizations. So in terms of their life spans, some are long and some are short. Now think about yourself, your birth, the beginning of your own life. Imagine that God does play dice, with simple probability rules. God creates you and randomly sends you—in accordance with these rules of probability—to live on some planet. Where will you be sent to live?

Imagine the civilizations as intervals of length—the length of time they have existed:

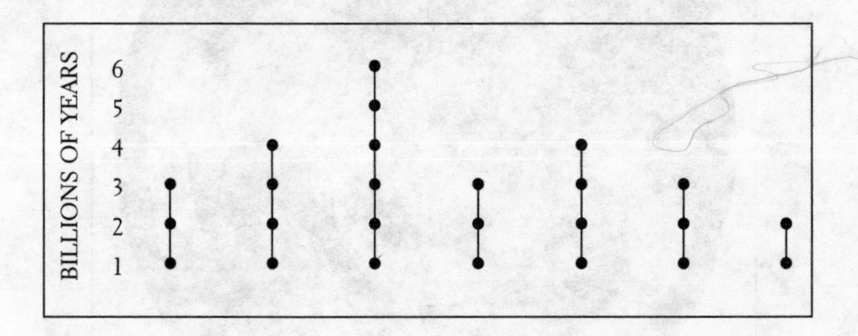

A longer-living civilization has a higher chance of receiving you than one that has existed for a short time. As with waiting for the bus, you are more likely to arrive in a world that has existed for a long time. But there is another factor here, which goes far beyond the simple inspection paradox.

If you arrived in a world that is young and has only produced bacteria, you would be a bacterium. And a bacterium is not interested in the probability of life in the universe. So the fact that we are advanced and intelligent *already* builds in an inspection bias. We inspect our own civilization, but our ability to inspect ourselves is an outcome of the fact that we've been here on this planet for a long time and have already evolved an intelligence that is able to ask such questions. The conditional probability that we have been around longer than other civilizations—conditional on the information that we are advanced enough to reason and wonder about life and the universe—is high.

Using these facts and the inspection paradox, we see that if your arrival on Earth is viewed as a random event, you are more likely to land on a longer-lived planet than on a shorter-lived one, in the same way that a dart is more likely to land on a wider section of the dartboard than on a narrower one.

By the inspection paradox, on the average you hit at the center point of an interval, and the interval you hit will have a larger-than-average length. In terms of the Sun, astronomers have explained that this is exactly what happened to us. The Sun is a star with a longer-than-average lifetime. Some stars, the more massive ones, live for millions of years and burn themselves out quickly, before life has a chance of developing. Other stars, like the Sun, live billions of years because they don't burn too bright and fast, since their mass is smaller. Our Sun has a total expected lifetime of over 10 billion years. And we've hit this longer-than-average interval right at its midpoint: the Sun is roughly 5 billion years old and is expected to last another 5 billion years. It is very likely that, as galactic civilizations go, we are on the above-average development level, and possibly way up there among the most advanced. Let's not forget other factors. In the early universe, there was not enough chemical diversity to allow for the development of life. Enough stars would have to have lived out their lifetimes and exploded in supernovas or shed off their atmospheres rich in chemical elements before there was enough richness in the primordial soup of gas and dust to form a protoplanetary disk that could spawn a life-giving planet like ours. This process takes time—billions of years. Our solar system is 5 billion years old, and the universe as we know it is roughly 14 billion years old—we may be one of the most advanced civilizations anywhere.

It appears that astronomers and the SETI people are

well aware of the implications of the inspection paradox even if they do not quite understand the theory. For astronomers are never interested in possible planets around young stars, bright stars such as Sirius, which burn out faster than the longer-lived Sun and other stars that burn for billions of years. Intuitively, these scientists guess that life may not have had a chance to develop on a planet that has been around for only millions of years.

Let's explore the fuller implications of the inspection paradox on life on Earth. We are here *now* because we needed all this time to get where we are—advanced beings with high intelligence. Just as an eighty-year-old woman can no longer die at age seventy and has an expected longevity much longer than the average woman, we can no longer be gorillas or lizards or algae. Earth required a billion and a half years for life to just begin here. Some scientists have assumed that the reason for the length of time is that DNA is a molecule so complicated that randomness alone would require a tremendous number of trials before it was formed by random fluctuations. But the reason may also involve the fact that early Earth did not have hospitable conditions for life. Violent geological activity on a fledgling planet—earthquakes and heavy volcanic activity, which filled the atmosphere with debris that prevented light from reaching the surface and caused unstable temperatures—and collisions with asteroids and comets in the newly formed solar system full of chaotically flying objects from creation may have made life here impossible for over a billion years. Lower life-forms have existed on Earth for 3.5 billion years, early animals for 700 million, land animals for 400 million, mammals for 200 million, the first primates for 100 million years. The gorilla evolved 10 million years ago; *Australopithecus,* 2 million years ago; *Homo habilis,* 1.5 million years ago; *Neanderthals,* 150,000 years ago; and *Homo sapiens* has been around for less than half a million years.

The ages of our planet are like those of a person whose life had begun long ago. The Sun is a middle-aged star, with half its life ahead of it, and our planet, therefore, is at its midpoint as well. But both are likely to be above average in total lifetime because of the inspection paradox. Therefore, life on other planets may well be at an earlier stage of development than ours. And this may explain the fact that we have not been overwhelmed by signs of life from outside our solar system. The time line on the next page shows how life evolved on Earth; it is a picture of the interval at which we are—the half-life of the total time interval of our solar system—and this interval is expected to be higher than the lifetime intervals of other civilizations.

The probability arguments above answer one important question: If extraterrestrial civilizations abound, why haven't we received any communications from them? This is the question often leveled at the SETI people. The answer is clear: Assuming other civilizations exist, chances are that we are among the first in our galaxy to arrive at this level of advancement. By the inspection paradox, we are probably more advanced than most other civilizations in the universe. Of course, there is also the question of why anyone should contact us by radio. Perhaps radio waves are not the preferred mode of communication of other civilizations. The SETI people themselves have been disheartened with the new developments of DirecTV and fiber-optic networks. If

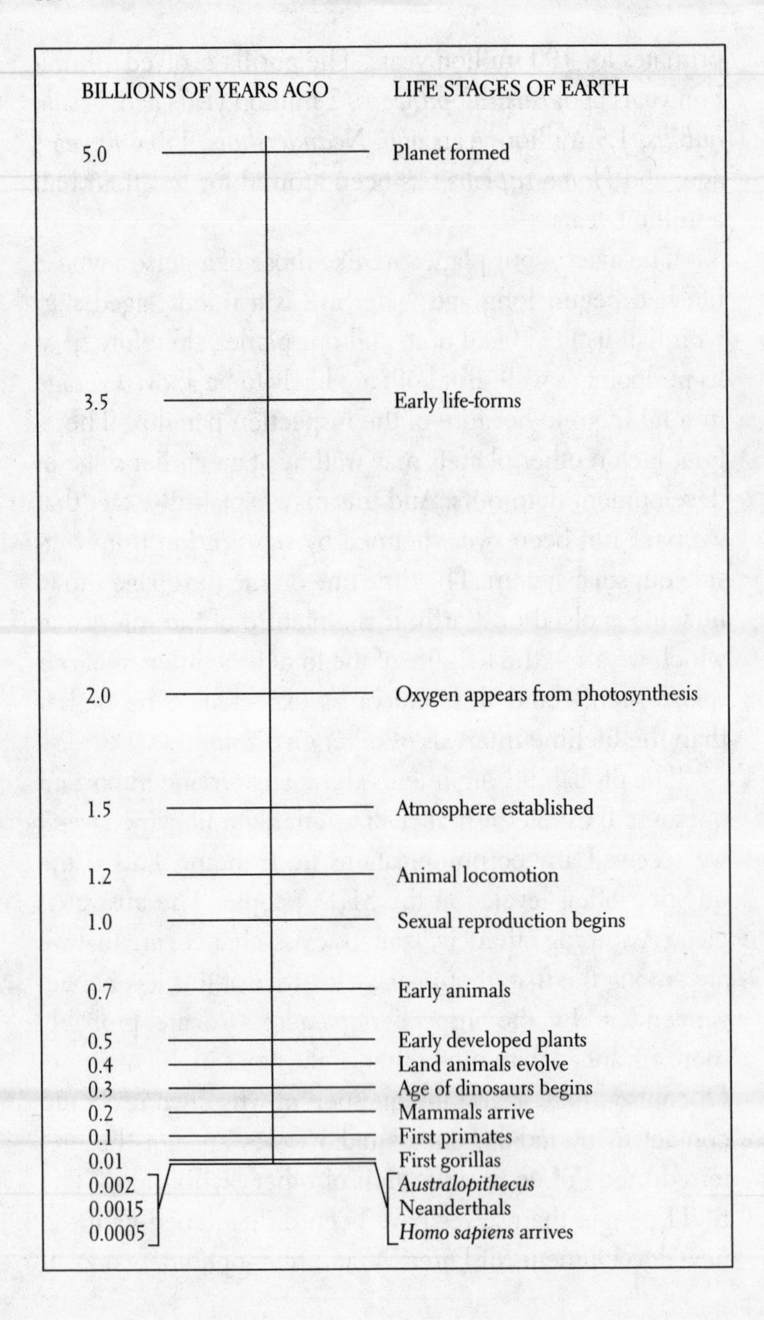
BILLIONS OF YEARS AGO
LIFE STAGES OF EARTH
5.0
Planet formed
3.5
Early life-forms
2.0
Oxygen appears from photosynthesis
1.5
Atmosphere established
1.2
Animal locomotion
1.0
Sexual reproduction begins
0.7
Early animals
0.5
Early developed plants
0.4
Land animals evolve
0.3
Age of dinosaurs begins
0.2
Mammals arrive
0.1
First primates
0.01
First gorillas
0.002
Australopithecus
0.0015
Neanderthals
0.0005
Homo sapiens arrives

our future communications look like this, and if others have found or will find that these modes are better than beaming radio waves indiscriminately around themselves, we will have little chance of forming radio contact. Possibly, establishing radio contact is not something on which we should concentrate our efforts. The tremendously large time delays inherent in such "communications" with even the closest stars (at least 4.25 years for a message to travel one way) make this an idea of dubious value.

In addition to explaining why we may very well be among the most advanced civilizations in the universe (assuming other civilizations exist) and hence our not receiving any information from other cultures in space, the inspection paradox does more. It exemplifies that in probability problems our intuition does not always work well. In the next two chapters we will encounter more unintuitive truths about probability and see how this powerful theory allows us to calculate the probability of the existence of life elsewhere in the cosmos.

10 The Birthday Problem

IN SOLVING AN INVOLVED probability problem, mathematicians have to build a pertinent model that will capture the essential elements of the problem. The model—a mental picture of what is going on—helps us understand the essence of the problem, and then, one hopes, it leads to the solution.

One model that has met with much success in the solution of probability problems is the balls-into-boxes analogy. This paradigm allows us to visually grasp the idea of matching, which is essential to many real-world probability problems. The most celebrated problem where the balls-into-boxes idea can be used successfully as a method of solution is the birthday problem. The solution of this problem is unexpected.

In a group of people in a room, what is the chance that at least two people share a common birthday? It becomes clear on first considering the problem that the answer depends on the number of people in the room; if there are

367 people, the probability of at least one match is 1.00, or 100 percent. This allows the possibility of a leap year, so the total number of possible dates is 366 (365 + February 29 on a leap year). Think of the days of the year as little boxes with open tops. There are 366 of them arranged in a row, and the people in the room are little balls falling randomly onto the line of open boxes. With 367 balls, even if each of the balls fell into a separate box (an extremely unlikely event, as we will see), the 367th ball would have to fall into a box already filled with another ball. At this point, a birthday is matched. Thus with 367 people, the probability of at least two people sharing a birthday is 100 percent.

But probabilities build up much quicker than this. Let's start with the people entering the room, each one giving his or her birthday date. The process is exactly like balls sequentially falling at random into the array of boxes below them. For simplicity's sake, let's assume there are 365 days in a year and ignore leap years. With this model, every one of the days of the year, 1/1 to 12/31, has an equal chance of receiving any single ball. Still, the outcome that each box will receive, at random, only one ball, is extremely small—that is, the chance of a match, a box with at least two balls in it, becomes quite high early on in the process of the balls falling down.

It turns out that with even 23 people in a room, the chance of at least one birthday match is *over* 50 percent. With 56 people in the room, the chance of at least two of them sharing a birthday in common is 99 percent! Even with only ten people in the room, the chance of a matched birthday is 12 percent. That is, with only ten balls falling down at random into a collection of 365 open boxes, the chance that the ten balls will occupy different boxes is only 88 percent (there is a 12 percent chance that at least two of the ten balls will fall into the same box).

Let's see mathematically why this happens the way it does. The first ball falling down has 365 random choices as to where to land. But if it is to miss the same box, the next ball has only 364 available boxes, so the probability that the second ball misses the box occupied by the first ball is $364/365$. If the third ball is to miss both boxes occupied by the first two balls, it will have to fall into one of 363 out of a total of 365 boxes. The process continues until all balls have fallen down. If there are 23 balls (23 people in the room), what is the chance that no two of them will fall into the same box (that is, all 23 people will have unique birthdays)? We need to multiply the possibilities for each of the balls because we want the joint outcome (ball 2 is in a different box from that of ball 1, *and* ball 3 is in a different box from that of the first two, and so on). We get:

Probability of 23 different boxes occupied by 23 different balls $= (365/365)(364/365)(363/365) \ldots (343/365)$
$= 0.4927.$

This can be easily verified using a simple calculator. Now, the probability of at least one matching birthday is equal to one minus the probability that all birthdays are unique (all 23 balls occupy different boxes). We subtract the answer from one because, for example, if the probability of rain is 30 percent, then the probability of no rain is 70 percent (and $0.7 = 1 - 0.3$). Thus the probability of at least one matching birthday among 23 people is equal to $1 - 0.4927$

= 0.5073, which is slightly over 50 percent. This result is, of course, surprising, for intuitively we don't expect the probability of a match to be so high when there are so many empty boxes still to be filled. And yet with only 23 balls, the chance of at least two of them in a box is already over 50 percent. Probability and intuition do not always agree. The mathematics always wins, with intuition or common sense sometimes on the losing side.

The balls-into-boxes analogy works very well in trying to figure out the probability that we are not alone in the cosmos. Let's pursue the analogy. But first, let's look at an extension of the balls-into-boxes idea, where the balls fall through the center of the line of boxes, with decreasing probabilities of falling away from the center. At the Museum of Science in Boston there is a glass display in the mathematics section. Here, balls fall from a central hole at the top of the display. The falling balls encounter an array of pegs at each level on their way down to the boxes below. At each peg, a ball will either fall left or right. Since each ball hits each peg straight on, it has a 50 percent chance of falling right and 50 percent chance of falling left at each level. As the balls fall down and pile up in the boxes below, a pattern emerges: the normal curve. In the center are balls that fell roughly the same number of times left as right. Farther and farther to the right are balls that fell more of the time to the right—and, by the fact that this has a small probability of occurring, there are fewer and fewer balls as you look farther to the right. The same pattern occurs on the left.

The display demonstrates the central limit theorem in action. This is the law of averages. On the average, at the

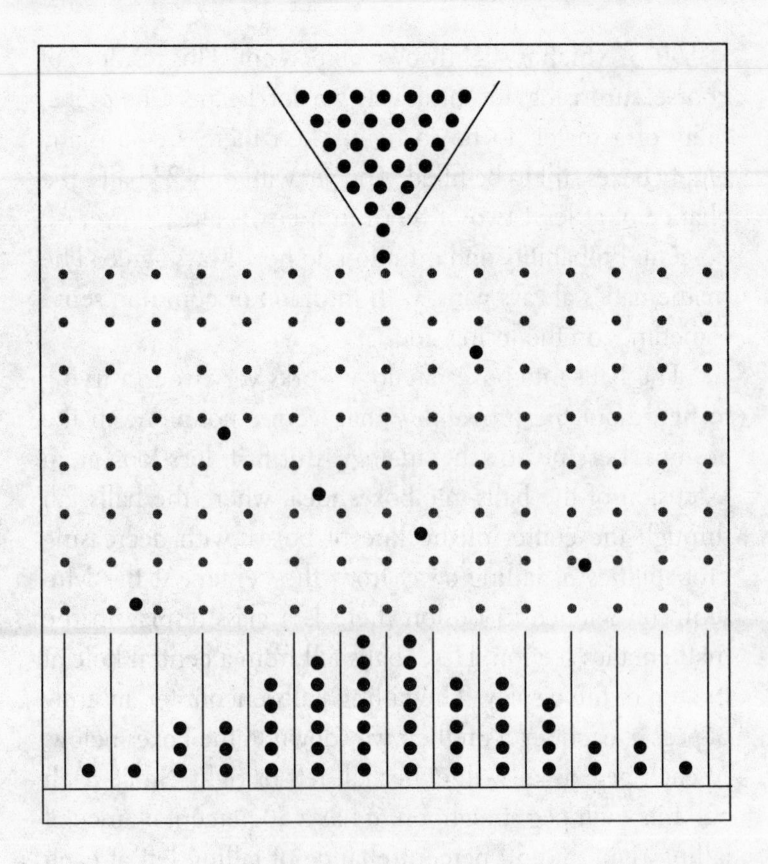

center, there are many balls. As you go away to the sides, there are fewer and fewer of them—and they are symmetrically distributed left and right since a sequence of six falls to the left and four to the right has the same probability as six falls to the right and four to the left. For us, what is interesting about the display is that even a rare event (a ball falling ten times to the left and zero to the right) does at times occur. And the normal curve (in the limit, as the number of pegs goes to infinity) assures us that rare events can occur. For by the normal probability law, there is a nonzero probability, small as it may be, that a value will occur as far to the left or as far to the right as one wishes.

Now let's consider the problem of life elsewhere in the universe. We already know from astronomy that extrasolar planets exist. We also know that some of them orbit their stars within the habitable zone, so life could evolve there. We also know, and scientists have known this for many decades, that the chemical compounds and elements that exist on Earth—including complex hydrocarbons and amino acids—also exist elsewhere in the known universe. What we don't know is whether the crucial compound of life, the DNA molecule, exists outside Earth.

Now let's look at the formation of the DNA molecule and assume it is constructed sequentially. Suppose we start with carbon from carbon dioxide, which exists everywhere in the universe, and methane, oxygen, hydrogen, elemental phosphorus, and sulfur—all known to exist everywhere in the known universe. At each stage, let's assume that one bond, out of the hundreds of chemical bonds that hold together the giant DNA molecule, is constructed. Let's assume that there is a small probability that the bond will occur when the elements involved are mixed together in a natural setting. These probabilities—one probability for each chemical bond—can be as small as we may wish. When a bond forms, it is just like the little ball falling down one level to the right in the glass display. Except that the probability of falling right (let's say) is very small, while that of falling left is very large. Still, as many balls fall down, some of them will eventually fall right. Then we reach the next level down—this is the next chemical bond in the DNA molecule being formed. The bond will be made if the ball falls right in this stage. Again, even if the probability of falling right is very small, still some balls will fall there. With a very large number of balls coming down, at least some of them will fall right every time, no matter how small the probability is of falling right at each single step.

Thus, with a tremendous number of trials (balls falling down)—which means a tremendously large combination of chemical soups on the surfaces of planets within their stars' habitable zones—and the right weather, temperature, pressure, and catalytic conditions, there will be a tiny but still nonzero probability that DNA will be formed. This is just like the fact that when many balls fall down the glass display, some of them will eventually land on the extreme right tail of the limiting normal distribution. How do we know this will happen? Well, the DNA molecule does exist, so its construction is possible. The only question here is how frequently it occurs: What is the probability that it will come into existence elsewhere? And here, our answer is that there is some small (even extremely small) but *nonzero* chance that it will occur on an extrasolar planet with an adequate wealth of chemical elements, ambient temperature, and so on.

The analogy of falling balls is even more pertinent for describing the formation of chemical bonds than meets the eye. For atoms are often viewed by chemists as little round balls. The chemical bonds—the sharing of electrons between atoms—are viewed as rigid wires at certain angles (120 degrees, 90 degrees, and so on) connecting the atoms. A bond has a given chance of occurring when two atoms hit each other at the right angle. From statistical mechanics, we know the probabilities of such reactions (two balls sticking together) at given temperature, pressure, and other conditions. And in statistical mechanics, the normal curve is the key to everything. Rates of reaction, and hence the frequency of occurrence of any type of bond, are all determined statistically for a large collection of atoms.

The frequency of occurrence of reactions in a large collection of atoms is exactly equal to the probability of occurrence of a *single* reaction between two atoms. From statistical mechanics we know that there are small, but nonzero, probabilities of occurrence of any complicated chemical bond. As these little balls hit against each other (or the pegs in the display analogy), they incur various probabilities of continuing to form more bonds, until the very complex DNA molecule appears. The problem is actually a little simpler than it seems. Some parts of the complex DNA molecule already exist as independent molecules. And we know from the findings of carbonaceous chondrites that even protein molecules exist outside of Earth. Once the submolecules—the bases adenine, guanine, cytosine, and thymine—exist, and the sugar molecule and the phosphate molecule that make up the backbone of DNA also exist outside of Earth, all it would take for DNA to be created is that these submolecules (rather than atoms—many of which would be required) hit against each other as little balls. If the reaction conditions are right, DNA will pop out. Conditional on the existence of the submolecules that make up DNA, the probability that it will materialize in space is greater than if these molecules were not known to exist as such.

Rates of chemical reaction have a strong connection with probability theory. The first question one asks in trying to assess a chemical reaction is whether the reaction can at all take place. Are the products of the chemical reaction stable? The answer to this question depends on thermodynamics. Are the energy levels such that the reactants will be able to produce its products? In the case of DNA, we already know the answer. Since the chemical compound does exist on Earth, we know for certain that the chemical reactions that produce the DNA molecule must be feasible, whatever form these reactions may take. This takes us quite a long way, for the only remaining questions are about the conditions that are required to produce this product. These are unknown to us, but we know that the right conditions must exist sometime, somewhere. Then the question is, what is the rate of the reaction?

Rates of chemical reaction depend on statistical factors. Atoms and molecules move around, their speeds depending directly on the temperature. As the temperature increases, the movements become faster. As the reactants move around, they collide with each other. Each time a collision occurs, there is a probability that a chemical bond will result and that the bond will hold together rather than break apart again. Some chemical reactions are very fast. When an acid meets a base in a water solution, the reaction is immediate: salt forms. Other reactions, such as the oxidation of iron to form rust, can take a long time. The rate of a chemical reaction depends on the solution of a differential equation. The equation and the process are governed by statistical probabilities. When two elements meet in a solution and the probability of the formation of a chemical bond as the result of collisions between the reactants is small, the rate of the reaction is low. Here, a lot of time is required for the many necessary collisions to take place since each time only a small fraction of these collisions result in a bond. (The probability of a bond forming from one collision is equal, in a large statistical system of many atoms, to the frequency of collisions that result in the bond.) When a reaction is fast, the probability of the formation of the bond from a single collision is high, and thus many bonds are formed in a period of time.

The reactions necessary for the formation of DNA are of course possible, but, from our experience on Earth, the reaction rate may be very, very slow. Some reactions take a fraction of a second to complete; others take several hours, while still others require a few days or weeks. Rust can form over a period of decades. The reactions necessary for DNA to form may take as long as a billion years. Over this length of time, many collisions among the submolecules that make up DNA occur, most of them unsuccessful like the balls that fall to the wrong side—but once in a great while, the correct sequence of occurrences takes place; the balls fall in the unlikely direction necessary for a rare reaction to take place.

Probability and time, as reflected by the rate of reaction, are strongly related. When the probability is very small (but nonzero), all it means is that a very large number of trials is necessary before a success. This is what happened when DNA was first created on Earth. The reaction took a very, very long time to materialize because the probability of the event was small. Our best estimate of how long these reactions would have required in order to occur elsewhere in space is the same as it took here, about a billion years or so. And this, again, agrees with the assumption that we probably are among the most advanced civilizations in our galaxy.

Let's go back to the original random balls example used in the explanation of the birthday problem, and let's view the situation from another angle. Assume that every element, and every condition—temperature, pressure, density, presence of a catalyst, and so on—is a little ball. These unique little balls fall randomly onto an array of open boxes below—without the central direction being favored. Here, just as in the original birthday problem, every little ball has an equal chance of falling into any one of the many boxes below. Instead of 365 boxes, as in the birthday problem, we have a large number. Each box represents an extrasolar planet within its star's habitable zone.

Each time a ball falls down into a box below, a condition for life has been realized. Thus, there will be stars with the right level of gravity, the right atmospheric pressure, the right temperature. The rest of the balls are chemical elements and compounds that are necessary for life. We know they exist here, so they must have a nonzero probability of existing in the universe beyond us. One ball falling down into a box below is the element carbon in sufficient abundance. This is a condition that probably is easy to satisfy, since we know that carbon is very abundant in the universe. Another ball is hydrogen; another is water molecules in sufficient quantities, and so on.

We know from the birthday problem that it is extremely *un*likely that the balls will all fall into different boxes—that is, that the conditions for life will be dispersed among different planets with no aggregation. The chance that all the necessary kinds of balls will fall into at least some of the many boxes is high, as long as there are enough balls for each of the conditions. If there are enough elemental particles in the universe, and science tells us that there are—there is no shortage of hydrogen, carbon, sulfur, oxygen and so on—then eventually at least some planets outside Earth would be showered with enough of the little balls supplying all of the ingredients necessary to allow life to evolve. In the next chapter we will finally compute the actual probability that life evolves outside Earth.

11 The Probability of Life on at Least One Other Planet

WE ARE NOW IN A POSITION to evaluate the final probability: the probability of life on at least one other planet in the universe. But in order to compute this aggregated probability that life exists in at least one other location in the cosmos, we must first develop the rule for computing the probabilities of such compound events. This is the rule for the union of a collection of independent events. It is one of the most important tools in the theory of probability, and it has an interesting history—for this rule, and much of the basic theory of probability, was developed because of the insatiable greed of a French gambler 350 years ago.

Blaise Pascal (1623–1662) was one of the most prominent French mathematicians of his day. He is well-known today for Pascal's law in physics, dealing with pressure in a system, and for his "Treatise on Conic Sections"—a single page he wrote when he was sixteen years old, which became the most famous single page in the history of mathematics, stating what we call Pascal's theorem. The theorem says that the opposite sides of a hexagon inscribed in a conic section intersect in three collinear points. From this result, the teenager went on to propose and prove a host of other theorems. When he was eighteen, Pascal invented a calculating machine—the forerunner of the computers made three hundred years later. (Pascal's calculating machine can be seen in the collection of the Arts and Sciences Department of IBM.)

In 1654, when he was thirty-one years old and one of the most famous mathematicians of his day, Blaise Pascal was visited by his friend the Chevalier de Méré. The Chevalier de Méré, Antoine Gombaud (1610–1685), was a wealthy French nobleman who was also an incorrigible gambler. De Méré was looking for ways to break the bank at the opulent casinos of Europe and wasn't shy about asking his famous mathematician friend for help in finding bets that would have a high probability of a win. Until then, nobody knew how to compute probabilities in a scientifically correct way. People knew that the probability of rolling a six on a die was $1/6$, but they did not know, for example, how to find the probability of rolling at least two sixes in twenty-four rolls of a pair of dice, which was a popular game at European casinos during that time. Another popular game was one where the gambler won if at least one six appeared in four rolls of a single die, and no one knew the probability of winning at this game, either.

De Méré thought that the probability of winning the first game was $24(1/36) = 2/3$ and that of winning the second game was $4(1/6) = 2/3$, and thus that both games offered the same chance of winning. But experience at the casinos showed that gamblers were winning more often at the second game, implying that the probability of getting at least one six in four rolls of a die was higher than the probability of getting at least two sixes in twenty-four rolls of a pair of dice. This apparent contradiction became known as the paradox of the Chevalier de Méré. The reason for this paradox was that de Méré did not know how to compute the probabilities of these compound events. The probability of at least one six in four throws of a die is not the same as four times the probability of getting a six in one throw of a die. What is needed here is a rule for the union of four events: the probability that a six happens at least once in four trials.[18]

So de Méré went to ask his friend Blaise Pascal why people were winning more often in the second game than they did in the first, and Pascal quickly realized that the gamblers did not know how to compute the probability of a union of events. Pascal thought about this problem for a while and discussed it with the mathematician Pierre de Fermat (1601–1665), who gave us Fermat's Last Theorem. Pascal and Fermat concluded that the probability of winning the first game was 49.1 percent, while for the second game the probability of winning was better than 50 percent: it was 51.8 percent. The discussions between Pascal and Fermat became the basis for the modern theory of probability.[19] How did the two mathematicians compute these probabilities, and what was the rule they derived for finding the probability of the union of several events?

Events that do not affect each other's probabilities are called independent events. The outcome of a die on one throw is independent of the outcome on another throw. Similarly, when we throw two dice, their outcomes are independent of each other. For independent events, we can multiply the probabilities to get the joint probability of occurrence. Another property is that the probability of the complement of an event is equal to one minus the probability of the event (thus if the probability of rain is 0.3, then the probability of no rain is $1 - 0.3 = 0.7$). Pascal and Fermat used these facts in deriving the product rule for the union of independent events. They reasoned that if at least one of several events occurs, then it is not the case that all of the events do not occur. So to get the probability of occurrence of at least one event, you subtract from one the probability that all the complements of the events will occur. But by independence, this latter probability—that the complements will all occur—is equal to the product of these probabilities. This gave them the final rule:

$$P(\text{at least one of } A, B, C \ldots) = 1 - P(\text{not } A) \times P(\text{not } B) \times P(\text{not } C) \ldots$$

In the case of the first game, betting on the occurrence of at least one double-six in twenty-four rolls of two dice, the union rule for independent events gave them the following answer:

$$1 - (35/36)(35/36)(35/36) \ldots (24 \text{ times}) = 1 - 0.509 = 0.491, \text{ or } 49.1\%$$

For the second game, Pascal and Fermat obtained the answer

$$1 - (5/6)(5/6)(5/6)(5/6) = 1 - 0.482 = 0.518, \text{ or } 51.8\%$$

Applying the Rule to the Probability of Extraterrestrial Life

The union rule for independent events allows us to compute the probability that there is *at least* one other planet outside Earth with life on it. Let's start by making some reasonable and minimal (that is, least favorable to our conclusion) assumptions about the basic probabilities of the existence of life on a planet orbiting any *one* star other than the Sun.

Let's take the estimate of the number of stars with planets, f_p from Drake's equation, as $f_p = 0.5$. Then, from the fact that out of nine extrasolar planets thus discovered, one is in the habitable zone, and the fact that this is confirmed in our own solar system (Earth being in the habitable zone, the other eight planets possibly not), we will use $\frac{1}{9}$ for that parameter. Now we come to the hard part, getting a lower bound for the actual probability of life: What is the probability that DNA develops and is sustained in life-forms on a planet that is within its star's habitable zone? Let's entertain the notion that DNA is an extremely complex molecule with a very small chance of occurring on its own and that life is precarious because the universe is a dangerous place. Let us therefore assume that the probability of life occurring on any single planet that is already within its star's habitable zone is extremely, extremely remote: *one in a trillion*. By multiplication of this extremely small number by the previous factors of 0.5 and $\frac{1}{9}$, we get the assumption that the probability of life around any one given star is 0.00000000000005.

Our galaxy has about 300 billion stars (although some estimates are lower), and let's assume there are 100 billion galaxies in the universe. We will now use all these estimates and plug them into the rule for the union of independent events:

$$P\text{ (life in orbit around at least one other star in the known universe)} = 1 - (0.99999999999995)^{30,000,000,000,000,000,000,000}$$

The answer is a number that is indistinguishable from 1.00 at any level of decimal accuracy reported by the computer. The answer is, for all practical purposes, equal to 1.00—or 100 percent.

Even if we assume that there are only 10 billion stars in our own galaxy and that there are only a billion galaxies, the answer still comes out to be a number indistinguishable from 1.00 for the probability of life elsewhere in the universe. This shows that the result is overwhelming—the probability that life exists outside Earth does not depend very strongly on the actual number of stars in the universe, as long as that number is very large—as we well know it to be from everything astronomy has taught us. New results from the Hubble Space Telescope about the existence of so many billions of galaxies in the universe serve the point that there are so many possible places for life to develop. There is also no dependence in the model on the assumptions about the percentage of stars with planets and the percentage of these planets within the habitable zone. While we used the best scientific estimates, even lower values still lead to the same answer, a number close to 1.00. The probability is a virtual certainty.

What is happening here, mathematically, is that even though our probability of life on any one planet may be extremely small, the compound probability that life exists on at least one other planet increases steadily because there are so many places to look—so many stars. This type of convergence as the number of trials becomes large always takes place when one uses the rule for the union of independent events. If you give something enough of a chance to happen, it eventually will.

Finally, we don't really know for certain the size of the entire universe. Some believe that our universe is infinite. If there are infinitely many stars, the answer to our question is that the probability of extraterrestrial life is *identically* equal to 1.00 (not just a number indistinguishable from 1.00 to any level of accuracy), and that this holds true *no matter how small the probability of life on any planet may be,* as long as that number is not identically zero (and we know that it is not zero since we exist).

Epilogue

In January 1998 Michel Mayor flew to Chile, to the large new observatory high in the Andes. A few months earlier, Paul Butler, the planet hunter from San Francisco State University, left for Australia. Both astronomers are now scanning the southern skies in search of more extrasolar planets. Their hope is to identify an Earth-sized planet orbiting a star like our Sun. Then, maybe life could be observed someday on such a planet.

The probability of extraterrestrial life is 1.00, or a number that for all purposes is 1.00. We are not alone. And while we haven't seen anyone from outside our planet yet, and while the distances to the stars are so dauntingly immense, someday in the future there might be contact. Courageous, insightful individuals in our history have taken strong positions against old, accepted opinions. Galileo argued that Earth was not the center of the universe and was punished by the Inquisition. Giordano Bruno, who in the sixteenth century had no access to astronomical observations of extrasolar planets or even to evidence that the stars are really distant suns, took his stand about the plurality of worlds and paid for it with his life. And in our own century, the story about the death of archaeologist Spyridon Marinatos demonstrates that it may still be dangerous to make far-reaching scientific conclusions in a world that resists new ideas.

I STOOD THERE, on the beach in Aruba, facing south, and gazed at the magnificent canopy of the night sky. The bright Canopus, beacon to generations of mariners, was slowly setting to my right. Straight above me I could see the three large constellations that make up Jason's Argo Navis, and it wasn't hard to imagine the great ship sailing into the Black Sea in search of the Golden Fleece. Farther left, the kitelike Southern Cross loomed above the horizon, and to my far left the stars of Alpha Centauri, our nearest neighbors, were rising as one. I wondered which of these stars had planets with life.

Endnotes

1. Epicurus, "Letter to Herodotus," in *The Stoic and Epicurean Philosophers*, ed. by W. J. Oates (New York: Random House, 1957).
2. Lucretius, *On the Nature of the Universe*, trans. by R. Latham (Baltimore: Penguin, 1951).
3. This is the *average* distance from Earth's orbit to the sun. Since the orbit is actually elliptical in shape, sometimes Earth is more than 93 million miles from the Sun and sometimes less. While it is elliptical, Earth's orbit is still close to a circle, and so the deviations from the average distance are small.
4. There are actually three stars in the system called Alpha Centauri, although to the naked eye they look like one star. The three orbit each other at relatively short distances. These are the dim star Proxima Centauri and the two brighter stars Alpha Centauri A and B.
5. Recent calculations imply that Hubble's constant is about sixty-four kilometers per second per megaparsec. The rate of the expansion of the universe is used by cosmologists

to estimate the elapsed time since the universe was created in the big bang. An expansion rate of sixty-four implies that our universe is about 14 billion years old.

6. Note that the gravitational effects of the other eight planets on our Sun's motion are much smaller than those of our largest planet, Jupiter, although the Sun's actual motion is slightly more complicated since all of these effects are present.
7. Sirrah is now considered to be in the constellation Andromeda. In the Middle Ages, Arabs were the world's most advanced astronomers, hence most star names today are still Arabic words.
8. This is an interesting misnomer. Astronomers who first observed the nebulae of gas and dust surrounding such dying stars thought they were observing dim, misty planets like Uranus, and named what they were seeing planetary nebulae.
9. Linus Pauling, *General Chemistry* (New York: Dover, 1988), p. 768.
10. I. S. Shklovskii and Carl Sagan, *Intelligent Life in the Universe* (San Francisco: Holden-Day 1966).
11. Sagan notes that both p and g are inversely proportional to the distance, r, of the organism from the Sun, and so is the value of the net force, $p - g$, propelling the organism.
12. Amir D. Aczel, "Improved Radiocarbon Age Estimation Using the Bootstrap," *Radiocarbon* 37, no. 3 (1995): 845–49.
13. Lewis Thomas, *The Lives of a Cell* (New York: Penguin 1973).
14. S. Dehaene, *The Number Sense* (Oxford: Oxford University Press, 1997).
15. For more on the fascinating story of the evolution of birds from dinosaurs, see K. Padian and L. Chiappe,

"The Origin of Birds and Their Flight," *Scientific American* (February 1998): 38–47.

16. It should be noted that chaos is not the only kind of system that is governed by mathematical equations—there are many other types of systems of equations. The simpler kind, linear equations (whose graphs are straight lines), are easy to detect. But nonlinear equations, including chaotic ones, are hard to discern but are much more prevalent in nature.
17. With twelve elements, six of each kind, 2, 3, 11, or 12 runs will be statistically significant for nonrandomness (using a level of significance close to 5 percent), while 4 to 10 runs could be assumed random since the randomness assumptions may not be rejected at this level.
18. By de Méré's logic, the probability of at least one head in two tosses of a coin would be 2(½) = 1, which we know cannot be true.
19. It should be noted that while we credit European mathematicians of the 1600s and 1700s with laying down the foundations of the modern theory of probability, we must recognize that some notions of probability had been well understood by people of many cultures around the world centuries earlier. For example, the great Indian epic *Mahabharata*, written before A.D. 400, discusses a demigod of dicing who can evaluate simple probabilities and even understands some ideas of statistical estimation. In the Talmud, written around the same time, rabbis discuss the probabilities that dietary laws are broken and probabilities of paternity and adultery—all based on simple rules that assign fifty-fifty chances to two equally uncertain events. See N. Rabinovitch, "Probability in the Talmud," *Biometrika* 56, no. 2 (1969): 437–41.

Further Reading

Aczel, Amir D. *Statistics: Concepts and Applications.* Burr Ridge, Ill.: Irwin/McGraw-Hill, 1993.

Boyer, C. *A History of Mathematics.* New York: Wiley, 1968.

Capra, F. *The Web of Life.* New York: Anchor, 1996.

Colata, Gina. *Clone.* New York: Morrow, 1998.

Crosswell, Ken. *Planet Quest.* New York: Free Press, 1997.

Crowe, M. *The Extraterrestrial Life Debate.* New York: Cambridge University Press, 1988.

Drake, Frank, and Dava Sobel. *Is Anyone Out There?* New York: Delacorte, 1992.

Encrenaz, T. *The Solar System.* New York: Springer-Verlag, 1990.

Freedman, D., et al. *Statistics.* New York: Norton, 1990.

Gehrels, T., ed. *Asteroids.* Tucson: University of Arizona Press, 1979.

Goldsmith, D. *Worlds Unnumbered.* Sausalito, Calif.: University Science Books, 1997.

Gould, S. J. *Ever Since Darwin.* New York: Norton, 1977.

Hoskin, M., ed. *The Illustrated History of Astronomy.* New York: Cambridge University Press, 1997.

Llinas, R. R., ed. *The Workings of the Brain.* New York: Freeman, 1990.

McDonough, T. *The Search for Extraterrestrial Intelligence.* New York: Wiley, 1987.

Murray, B., et al. *Earthlike Planets.* New York: Freeman, 1981.

Pasachoff, Jay M. Astronomy: From the Earth to the Universe, Fifth Edition, Ft. Worth: Saunders, 1998.

Pauling, Linus. *General Chemistry.* New York: Dover, 1988.

Sagan, Carl, ed. *Communications with Extraterrestrial Intelligence.* Cambridge, Mass.: MIT Press, 1973.

Shklovskii, I. S., and Carl Sagan. *Intelligent Life in the Universe.* San Francisco: Holden-Day, 1966.

Thomas, Lewis. *The Lives of a Cell.* New York: Penguin, 1973.

Watson, James D. *The Double Helix.* New York: Penguin, 1968.

Index